EDUCATION IN A COMPETITIVE AND GLOBALIZING WORLD

U.S. SCIENCE, TECHNOLOGY, ENGINEERING AND MATH (STEM) EDUCATION

EDUCATION IN A COMPETITIVE AND GLOBALIZING WORLD

Additional books in this series can be found on Nova's website under the Series tab.

Additional E-books in this series can be found on Nova's website under the E-books tab.

EDUCATION IN A COMPETITIVE AND GLOBALIZING WORLD

U.S. SCIENCE, TECHNOLOGY, ENGINEERING AND MATH (STEM) EDUCATION

JASON C. ROLLINS
EDITOR

Nova Science Publishers, Inc.
New York

For permission to use material from this book please contact us:
Telephone 631-231-7269; Fax 631-231-8175
Web Site: http://www.novapublishers.com

Library of Congress Cataloging-in-Publication Data

U.S. science, technology, engineering and math (STEM) education / [edited by] Jason C. Rollins.
p. cm.
Includes index.
ISBN 978-1-61209-792-3 (hardcover)
1. Science--Study and teaching--United States. 2. Technology--Study and teaching--United States. 3. Mathematics--Study and teaching--United States. I. Rollins, Jason C. II. President's Council of Advisors on Science and Technology (U.S.) III. National Science Board (U.S.)
Q183.3.A1U158 2011
507.1'073--dc22
2011003978

Published by Nova Science Publishers, Inc. † New York

CONTENTS

PREFACE

The success of the United States in the 21st century, its wealth and welfare, will depend on the ideas and skills of its population. As the world becomes increasingly technological, the value of these national assets will be determined in no small measure by the effectiveness of science, technology, engineering and mathematics (STEM) education in the United States. STEM education will determine whether the United States will remain a leader among nations and whether we will be able to solve immense challenges in such areas as energy, health, environmental protection, and national security. This new book explores a strategy for improving K-12 STEM education that responds to the tremendous challenges and historic opportunities facing the Nation.

Chapter 1- The success of the United States in the 21st century – its wealth and welfare – will depend on the ideas and skills of its population. These have always been the Nation's most important assets. As the world becomes increasingly technological, the value of these national assets will be determined in no small measure by the effectiveness of science, technology, engineering, and mathematics (STEM) education in the United States. STEM education will determine whether the United States will remain a leader among nations and whether we will be able to solve immense challenges in such areas as energy, health, environmental protection, and national security. It will help produce the capable and flexible workforce needed to compete in a global marketplace. It will ensure our society continues to make fundamental discoveries and to advance our understanding of ourselves, our planet, and the universe. It will generate the scientists, technologists, engineers, and mathematicians who will create the new ideas, new products, and entirely new industries of the 21st century. It will provide the technical skills and quantitative literacy needed for individuals to earn livable wages and make better decisions for themselves, their families, and their communities. And it will strengthen our democracy by preparing all citizens to make informed choices in an increasingly technological world.

Chapter 2- A little more than a decade later, mobilized by the Soviet's successful launch of Sputnik, the United States embarked on a collective, coordinated, and sustained effort to recruit and educate the "best and brightest" who subsequently would form a new generation of leaders and innovators in science and engineering. This effort resulted in unprecedented scientific and technological progress, leading to the creation of new enterprises, new jobs, and the betterment of the national standard of living. At the root of this progress was a substantial investment in research and development, along with a nationwide focus on excellence in science, technology, engineering, and mathematics (STEM) education and talent development. Regrettably, by the 1970s, this national sense of urgency had diminished, and complacency soon supplanted the ideal of excellence in education. Today, faced with growing international competition, the cost of inaction continues to grow at an intensifying pace.

In: U.S. Science, Technology, Engineering and Math (STEM)… ISBN: 978-1-61209-792-3
Editor: Jason C. Rollins

Chapter 1

Report to the President: Prepare and Inspire: K-12 Education in Science, Technology, Engineering and Math (STEM) for America's Future

President's Council of Advisors on Science and Technology

Executive Report

The success of the United States in the 21st century – its wealth and welfare – will depend on the ideas and skills of its population. These have always been the Nation's most important assets. As the world becomes increasingly technological, the value of these national assets will be determined in no small measure by the effectiveness of science, technology, engineering, and mathematics (STEM) education in the United States. STEM education will determine whether the United States will remain a leader among nations and whether we will be able to solve immense challenges in such areas as energy, health, environmental protection, and national security. It will help produce the capable and flexible workforce needed to compete in a global marketplace. It will ensure our society continues to make fundamental discoveries and to advance our understanding of ourselves, our planet, and the universe. It will generate the scientists, technologists, engineers, and mathematicians who will create the new ideas, new products, and entirely new industries of the 21st century. It will provide the technical skills and quantitative literacy needed for individuals to earn livable wages and make better decisions for themselves, their families, and their communities. And it will strengthen our democracy by preparing all citizens to make informed choices in an increasingly technological world.

Throughout the 20th century, the U.S. education system drove much of our Nation's economic growth and prosperity. The great expansion of high school education early in the century, followed by an unprecedented expansion of higher education, produced workers with high levels of technical skills, which supported the economy's prodigious growth and reduced economic inequality. At the same time, scientific progress became an increasingly important driver of innovation-based growth. Since the beginning of the 20th century, average per capita income in the United States has grown more than sevenfold, and science and technology account for more than half of this growth. In the 21st century, the country's need for a world-leading STEM workforce and a scientifically, mathematically, and technologically literate populace has become even greater, and it will continue to grow – particularly as other nations continue to make rapid advances in science and technology. In the words of President Obama, "We must educate our children to compete in an age where knowledge is capital, and the marketplace is global."

Troubling Signs

Despite our historical record of achievement, the United States now lags behind other nations in STEM education at the elementary and secondary levels. International comparisons of our students' performance in science and mathematics consistently place the United States in the middle of the pack or lower. On the National Assessment of Educational Progress, less than one-third of U.S. eighth graders show proficiency in mathematics and science.

Moreover, there is a large interest and achievement gap among some groups in STEM, and African Americans, Hispanics, Native Americans, and women are seriously underrepresented in many STEM fields. This limits their participation in many well-paid, high-growth professions and deprives the Nation of the full benefit of their talents and perspectives.

It is important to note that the problem is not just a lack of *proficiency* among American students; there is also a lack of *interest* in STEM fields among many students. Recent evidence suggests that many of the most proficient students, including minority students and women, have been gravitating away from science and engineering toward other professions. Even as the United States focuses on low-performing students, we must devote considerable attention and resources to all of our most high-achieving students from across all groups.

What lies behind mediocre test scores and the pervasive lack of interest in STEM is also troubling. Some of the problem, to be sure, is attributable to schools that are failing systemically; this aspect of the problem must be addressed with systemic solutions. Yet even schools that are generally successful often fall short in STEM fields. Schools often lack teachers who know how to teach science and mathematics effectively, and who know and love their subject well enough to inspire their students. Teachers lack adequate support, including appropriate professional development as well as interesting and intriguing curricula. School systems lack tools for assessing progress and rewarding success. The Nation lacks clear, shared standards for science and math that would help all actors in the system set and achieve goals. As a result, too many American students conclude early in their education that STEM subjects are boring, too difficult, or unwelcoming, leaving them ill-prepared to meet the challenges that will face their generation, their country, and the world.

National Assets and Recent Progress

Despite these troubling signs, the Nation has great strengths on which it can draw.

First, the United States has the most vibrant and productive STEM community in the world, extending from our colleges and universities to our start-up and large companies to our science-rich institutions such as museums and science centers. The approximately 20 million people in the United States who have degrees in STEM- or healthcare-related fields can potentially be a tremendous asset to U.S. education.

Second, a growing body of research has illuminated how children learn about STEM, making it possible to devise more effective instructional materials and teaching strategies. The National Research Council and other organizations have summarized this research in a number of influential reports and have drawn on it to make recommendations concerning the teaching of mathematics and science. These reports transcend tired debates about conceptual understanding versus factual recall versus procedural fluency. They emphasize that students learning science and mathematics need to acquire all of these capabilities, because they support each other.

Third, a clear bipartisan consensus has emerged on the need for education reform in general and the importance of STEM education in particular. The 2002 reauthorization of the Elementary and Secondary Education Act, renamed the No Child Left Behind Act, established the importance of collecting data annually about students' and schools' progress in mathematics and reading and tied Federal education funding to progress. The Congress is currently working on reauthorization of this law, with modifications to improve it.

The Obama administration has made education reform one of its highest priorities. The American Recovery and Reinvestment Act of 2009 established four broad "assurances" to improve the K-12 education system, and the administration has worked to fulfill these assurances through competitive grant-making. A historic, state-led initiative – led by the National Governors Association and the Council of Chief State School Officers – emerged in 2008 to forge clear, consistent, and higher standards for mathematics and English language arts education in grades K-12 that can be shared across states. These standards were recently released, and, as of the publication date of this report, 36 states and the District of Columbia had adopted them. There is also considerable interest in the adoption of similar standards for science, which will be essential for improving STEM education.

Purpose of this Report

In the fall of 2009, the President asked his President's Council of Advisors on Science and Technology (PCAST) to develop specific recommendations concerning the most important actions that the administration should take to ensure that the United States is a leader in STEM education in the coming decades. In responding to this charge, PCAST decided to focus initially on the K-12 level. (A subsequent report will address STEM education at community colleges, four-year colleges, and universities.)

There have been a number of important reports related to STEM education over the past two decades, including landmark reports that have called attention to the problem, reviews of the research literature, and recommendations concerning principles and priorities. Our goal is not to redo the work of these excellent reports – indeed, we have relied heavily on their research and findings. Rather, the purpose of this PCAST report is instead to translate these ideas into a coherent program of Federal action to support STEM education in the United States that responds to current opportunities.

The report examines the national goals and necessary strategies for successful STEM education. We examine the history of Federal support for STEM education and consider actions that the Federal Government should take with respect to improving leadership and coordination. Subsequent chapters discuss Standards and Assessments, Teachers, Technology, Students and Schools.

Many of the recommendations in this report can be carried out with existing Federal funding. Some of the recommendations could be funded in part through existing programs, although new authorities may be required in certain cases. Depending on these choices, the new funding required to fully fund the recommendations could reach up to approximately $1 billion per year. This would correspond to the equivalent of roughly $20 per K-12 public school student; or 2 percent of the total Federal spending of approximately $47 billion on K-12 education; or 0.17 percent of the Nation's total spending of approximately $593 billion on K-12 education. Not all of this funding must come from the Federal budget. We believe that some of the funding can come from private foundations and corporations, as well as from states and districts.

Key Conclusions and Recommendations

While the report discusses a range of conclusions and recommendations, we have sought to identify the most critical priorities for rapid action. Below, we summarize our two main conclusions and our seven highest priority recommendations.

All of these recommendations are directed at the Federal Government, and in particular we focus our attention on actions to be taken by the Department of Education and the National Science Foundation as the lead Federal agencies for STEM education initiatives in K-12.

Achieving the Nation's goals for STEM education in K-12 will require partnerships with state and local government and with the private and philanthropic sectors. The Federal Government must actively engage with each of these partners, who must in turn fulfill their own distinctive roles and responsibilities. In this context, we are encouraged by the state-led collaborative efforts and by the creation of private groups, such as the recently formed coalition, Change the Equation.

CONCLUSIONS

To Improve Stem Education, We Must Focus on Both Preparation and Inspiration

To meet our needs for a STEM-capable citizenry, a STEM-proficient workforce, and future STEM experts, the Nation must focus on two complementary goals: We must *prepare* all students, including girls and minorities who are underrepresented in these fields, to be proficient in STEM subjects. And we must *inspire* all students to learn STEM and, in the process, motivate many of them to pursue STEM careers.

The Federal Government Has Historically Lacked a Coherent Strategy and Sufficient Leadership Capacity for K-12 Stem Education

Over the past few decades, a diversity of Federal projects and approaches to K-12 STEM education across multiple agencies appears to have emerged largely without a coherent vision and without careful oversight of goals and outcomes. In addition, relatively little Federal funding has historically been targeted toward catalytic efforts with the potential to transform STEM education, too little attention has been paid to replication and scale-up to disseminate proven programs widely, and too little capacity at key agencies has been devoted to strategy and coordination.

RECOMMENDATIONS

(1) Standards: Support the Current State-Led Movement for Shared Standards in Math and Science

The Federal Government should vigorously support the state-led effort to develop common standards in STEM subjects, by providing financial and technical support to states for (i) rigorous, high-quality professional development aligned with shared standards, and (ii) the development, evaluation, administration, and ongoing improvement of assessments aligned to those standards.

The standards and assessments should reflect the mix of factual knowledge, conceptual understanding, procedural skills, and habits of thought described in recent studies by the National Research Council.

(2) Teachers: Recruit and Train 100,000 Great Stem Teachers over the Next Decade Who are Able to Prepare and Inspire Students

The most important factor in ensuring excellence is great STEM teachers, with both deep content knowledge in STEM subjects and mastery of the pedagogical skills required to teach these subjects well.

The Federal Government should set a goal of ensuring over the next decade the recruitment, preparation, and induction support of at least 100,000 new STEM middle and high school teachers who have strong majors in STEM fields and strong content-specific pedagogical preparation, by providing vigorous support for programs designed to produce such teachers.

(3) Teachers: Recognize and Reward the Top 5 Percent of the Nation's Stem Teachers, by Creating a Stem Master Teachers Corps

Attracting and retaining great STEM teachers requires recognizing and rewarding excellence.

The Federal Government should support the creation of a national STEM Master Teachers Corps that recognizes, rewards, and engages the best STEM teachers and elevates the status of the profession. It should recognize the top 5 percent of all STEM teachers in the Nation, and Corps members should receive significant salary supplements as well as funds to support activities in their schools and districts.

(4) Educational Technology: Use Technology to Drive Innovation, By Creating an Advanced Research Projects Agency for Education

Information and computation technology can be a powerful driving force for innovation in education, by improving the quality of instructional materials available to teachers and students, aiding in the development of high-quality assessments that capture student learning, and accelerating the collection and use of data to provide rich feedback to students, teachers, and schools. Moreover, technology has been advancing rapidly to the point that it can soon play a transformative role in education.

Realizing the benefits of technology for K-12 education, however, will require active investments in research and development to create broadly useful technology platforms and well-designed and validated examples of comprehensive, integrated "deeply digital" instructional materials.

The Federal Government should create a mission-driven, advanced research projects agency for education (ARPA-ED) housed either in the Department of Education, in the National Science Foundation, or as a joint entity. It should have a mission-driven culture, visionary leadership, and draw on the strengths of both agencies. ARPA-ED should propel and support (i) the development of innovative technologies and technology platforms for learning, teaching, and assessment across all subjects and ages, and (ii) the development of effective, integrated, whole-course materials for STEM education.

(5) Students: Create Opportunities for Inspiration through Individual and Group Experiences Outside the Classroom

STEM education is most successful when students develop personal connections with the ideas and excitement of STEM fields. This can occur not only in the classroom but also through individualized and group experiences outside the classroom and through advanced courses.

The Federal Government should develop a coordinated initiative, which we call INSPIRE, to support the development of a wide range of high-quality STEM-based after-school and extended day activities (such as STEM contests, fabrication laboratories, summer and afterschool programs, and similar activities). The program should span disparate efforts of science mission agencies and after-school programs supported through the Department of Education funding.

(6) Schools: Create 1,000 New Stem-Focused Schools over the Next Decade

STEM-focused schools represent a unique National resource, both through their direct impact on students and as laboratories for experimenting with innovative approaches. The Nation currently has only about 100 STEM-focused schools, concentrated at the high school level.

The Federal Government should promote the creation of at least 200 new highly-STEM-focused high schools and 800 STEM-focused elementary and middle schools over the next decade, including many serving minority and high-poverty communities. In addition, the Federal Government should take steps to ensure that all schools and schools systems have access to relevant STEM-expertise.

(7) Ensure Strong and Strategic National Leadership

Stronger leadership, coherent strategy and greater coordination are essential to support innovation in K-12 STEM education. Toward this end, the Federal Government should (i) create new mechanisms, with substantially increased capacity, to provide leadership within each of the Department of Education and the National Science Foundation; (ii) establish a high-level partnership between these agencies; (iii) establish a standing Committee on STEM Education within the National Science and Technology Council responsible for creating a Federal STEM education strategy; and (iv) establish an independent Presidential Commission on STEM Education, in conjunction with the National Governors Association, to promote and monitor progress toward improving STEM education.

PCAST believes that the Nation has an urgent need – but also, thanks to recent developments, an unprecedented opportunity – to bring together stakeholders at all levels to transform STEM education to lay the groundwork for a new century of American progress and prosperity.

I. INTRODUCTION AND CHARGE

Chapter Summary

The Nation's future depends on our ability to educate today's students in science, technology, engineering, and mathematics (STEM). Despite the fact that many U.S. students excel in STEM, U.S. students as a whole perform poorly on international comparisons of mathematical and scientific proficiency. There are wide disparities in STEM achievement among groups, and too many students think of STEM subjects as too difficult or uninviting. Nevertheless, the Nation can draw on key strengths to address these challenges, including a large and vibrant community of STEM professionals, new understandings of how children learn, a bipartisan consensus about the importance of STEM education, and state-led movements toward agreement on what students should learn in STEM. We must seize this historic moment by making changes and investments to educate all students for a future in which science and technology will play a critical role in the lives of individuals and the prospects of nations.

Introduction

The success of the United States in the 21st century – its wealth and welfare – will depend on the ideas and skills of its population. These have always been the Nation's most important assets. As the world becomes increasingly technological, the value of these national assets will be determined in no small measure by the effectiveness of science, technology, engineering, and mathematics (STEM) education in the United States.

STEM education will determine whether the United States will remain a leader among nations and whether we will be able to solve immense challenges in such areas as energy, health, environmental protection, and national security. It will help produce the capable and flexible workforce needed to compete in a global marketplace. It will ensure our society continues to make fundamental discoveries and to advance our understanding of ourselves, our planet, and the universe. It will generate the scientists, technologists, engineers, and mathematicians who will create the new ideas, new products, and entirely new industries of the 21st century. It will provide the technical skills and quantitative literacy needed for individuals to earn livable wages and make better decisions for themselves, their families, and their communities. And it will strengthen our democracy by preparing all citizens to make informed choices in an increasingly technological world. Given its importance, STEM education must prepare and engage all students no matter their gender, race, or background.

Throughout the 20th century, the U.S. education system drove much of our Nation's economic growth and prosperity.[1] The great expansion of high school education early in the century, followed by an unprecedented expansion of higher education, produced workers with high levels of technical skills, which supported the economy's prodigious growth and reduced economic inequality. At the same time, scientific progress became an increasingly important driver of innovation-based growth.[2] Since the beginning of the 20th century, average per capita income in the United States has grown more than sevenfold,[3] and science and technology account for more than half of this growth.[4] The fastest growing occupations in the United States are in healthcare and social assistance and professional, scientific, and technical services.[5] Inventions in which America played a central role, such as the airplane, the television, the computer, the Internet, and biotechnology, have changed the world.

In the 21st century, the country's need for a world-leading STEM workforce and a scientifically, mathematically, and technologically literate populace has become even greater, and it will continue to grow – particularly as other nations continue to make rapid advances in science and technology. In the words of President Obama, "We must educate our children to compete in an age where knowledge is capital, and the marketplace is global." STEM education is essential to our economic competitiveness and our national, health, and environmental security. It is also our obligation to empower future generations with the tools and knowledge they will need to seize the opportunities and solve the global problems that they will inherit. STEM education is critical to the Nation's roles and responsibilities in the world, including our ability to play a role in international development.

Troubling Signs

Despite our historical record of achievement, the United States now lags behind other nations in STEM education at the elementary and secondary levels. Over the past several decades, a variety of indicators have made clear that we are failing to educate many of our young people to compete in an increasingly high-tech global economy and to contribute to national goals.

International comparisons of our students' performance in science and mathematics place the United States in the middle of the pack or lower. The Trends in International Mathematics and Science Study (TIMSS) puts U.S. fourth graders and eighth graders about average among industrialized and rapidly industrializing countries.[6] However, U.S. students in fourth, eighth, and twelfth grades drop progressively lower on international comparisons of science and mathematics ability as their grade level increases. Also, in the Programme for International Student Assessment (PISA), which measures students' ability to apply what they have learned in science and technology and has been designed to assess the kinds of skills needed in today's workplace, U.S. 15-year-olds scored below most other nations tested in 2006, and the U.S. standing dropped from 2000 to 2006 in both math and science.[7] On the National Assessment of Educational Progress (NAEP), less than one-third of U.S. eighth graders show proficiency in mathematics and science, and science test scores have improved very little over the past few decades. This is not an acceptable standard of achievement for our Nation.

This inadequate preparation in STEM subjects has major consequences in higher education. Only about a third of bachelor's degrees earned in the United States are in a STEM field, compared with approximately 53 percent of first university degrees earned in China, and 63 percent of those earned in Japan.[8] More than half of the science and engineering graduate students in U.S. universities are from outside the United States. It is good for the Nation that our universities are a beacon to the world's best students: many of these students stay and contribute to the growth of our economy, while others return home with knowledge of and ties to this country. But it is troubling that the proportion of Americans interested in such graduate study is so low.

Moreover, there is a large interest and achievement gap in the United States in STEM. As a result, African Americans, Hispanics, Native Americans, and women are seriously underrepresented in many STEM fields, which limits their participation in many well-paid, high-growth professions. The underrepresentation of minority groups and women in STEM denies the Nation the full benefit of their talents and denies science and engineering the rich diversity of perspectives and inspiration that drive those fields. Diversity is essential to producing scientific innovation, and we cannot solve the STEM crisis the country faces without improving STEM achievement across gender and ethnic groups.[9] Moreover, all students deserve the opportunity to experience the exciting and inspiring aspects of STEM.

It is important to note that the problem is not just a lack of *proficiency* among American students; there is also a lack of *interest* in STEM fields among many students. The United States has historically benefited when talented and high-achieving students have entered STEM fields. But recent evidence suggests that many of these students, including minority students and women, have been gravitating away from science and engineering toward other professions.[10] A gender gap persists not in STEM aptitude but in interest: Although girls earn high school mathematics and science credits at the same rate as boys, and earn slightly higher grades in those classes,[11] they choose STEM majors in college at a much lower rate than boys.[12] Girls who are high achievers in mathematics in the United States are concentrated at a small number of high schools, which suggests that most girls with high ability to excel in the field are not doing so.[13] Even as the United States focuses on low-performing students, we must devote considerable attention and resources to all of our most high-achieving and high-ability students from across all groups.

There are some bright spots with respect to student performance and interest in STEM subjects. Math test scores at the fourth and eighth grade levels have increased over the past two decades,[14] at least in part due to higher standards and greater accountability. On the TIMSS exam, the United States' standing among comparison nations rose slightly from 1995 to 2007 in mathematics (but not in science). Some of the achievement gaps between groups of students have narrowed. For example, Hispanic and African American students increased their mathematical performance between 2000 and 2007 and narrowed the gap with white students. [15] Some individual states also perform at relatively high levels. In Massachusetts, fourth graders score behind only two jurisdictions in math (Hong Kong and Singapore) and behind only one jurisdiction in science (Singapore).[16] In Minnesota, the scores are only slightly lower. There are hints that participation in some STEM courses has increased; since the late 1980s, the proportion of public high school seniors who graduate having taken at least one physics course has risen from less than 20 percent to 37 percent.[17] These results demonstrate that positive movement is possible, but progress has been slow and often slight, and it is not sufficient to get all U.S. students – regardless of where they live – to where they need to be.

What lies behind our mediocre test scores and lack of interest is also troubling. Some of the problem, to be sure, is attributable to schools that are failing systemically; this aspect of the problem must be addressed with systemic solutions. Yet even schools that are generally successful often fall short in STEM fields. Schools often lack teachers who know how to teach science and mathematics effectively, and who know and love their subject well enough to inspire their students. Teachers lack adequate support, including appropriate professional development as well as interesting and intriguing curricula. School systems lack tools for assessing progress and rewarding success. The Nation lacks clear, shared standards for STEM subjects that would help all actors in the system set and achieve goals. As a result, too many American students conclude early in their education that STEM subjects are boring, too difficult, or unwelcoming, leaving them ill-prepared to meet the challenges that will face their generation, their country, and the world.

National Assets and Recent Progress

To meet these challenges, the United States has great strengths on which it can draw, including the world's leading community of scientists, technologists, engineers, and mathematicians. In addition, important progress has recently been made in understanding how to improve STEM education and in developing a national consensus about how best to move forward.

1. The U.S. STEM Professional Community. The United States has the most vibrant and productive STEM community in the world, extending from our colleges and universities to our start-up and large companies to our science-rich institutions such as museums and science centers. U.S. colleges and universities continue to attract many of the world's brightest and most dedicated students. Many of these foreign students join the U.S. workforce and make major contributions to our Nation's economy and culture. Since the 1950s, Americans have won more Nobel Prizes in science than scientists of any other nationality (though this is a lagging indicator, reflecting past accomplishments rather than current educational excellence).

The approximately 20 million people in the United States who have degrees in STEM fields or healthcare can potentially be a tremendous asset to U.S. education.[18] The leadership of the STEM community is engaged in policy discussions and is eager to improve STEM education. Moreover, a great many scientists and engineers would be willing to contribute to improving STEM education, both in school and out of school, if an efficient and effective way for them to do so could be put in place. In particular, since scientists and engineers are already well versed in the use of information technologies, web-based mechanisms that facilitate such contributions should be maximized.

2. Research Progress. A growing body of research in recent decades has illuminated how children learn about science, math, and technology, which is making it possible to devise more effective instructional materials and teaching strategies. This progress has been summarized in influential reports by the National Research Council and other organizations.[19,20] For example, studies have pointed toward the effectiveness of "active learning," which occurs when children are interacting with teachers, classmates, and environments or undertaking projects rather than passively taking in whatever a teacher tells them. Research also suggests that trying to cover too many topics in a curriculum with too little in-depth study can impair conceptual understanding.[21, 22, 23] Research on "learning progressions" – which describe the hierarchical understandings children obtain in science and mathematics – also has made considerable progress; it points toward the concepts that all children must acquire and highlights common difficulties students face that hinder learning.[24] Studies have also emerged showing that learning occurs everywhere and that a learner's waking hours outside of school can be critically important to STEM learning and interest.[25]

Furthermore, studies suggest that achieving expertise is less a matter of innate talent than of having the opportunity and motivation to dedicate oneself to the study of a subject in a productive, intellectual way – and for sufficient time – to enable the brain development needed to think like a scientist, mathematician, or engineer. This has important implications for STEM education; it underscores the need to motivate students for long-term study of STEM, and points to the potential for many more students to excel in STEM.[26, 27, 28]

Put together, this body of evidence suggests that grade-school children do not think as simplistically about STEM subjects as conventional curricula assume. They are capable of grasping both concrete examples and abstract concepts at remarkably early ages. Conventional approaches to teaching science and math have sometimes been shaped by misconceptions about what children cannot learn rather than focusing on students' innate curiosity, reasoning skills, and intimate observations of the natural world. STEM educators and standard-setters can now draw on such knowledge to design curricula that are age-appropriate and engage students in observing the world, testing what they find against what they expect, and teaching each other.

The National Research Council has drawn on this research to make recommendations concerning the teaching of mathematics[29] and science.[30] These reports transcend the tired debates about conceptual understanding versus factual recall versus procedural fluency. They emphasize that students learning science and mathematics need to master all of these capabilities, because they support each other. Math education should, for example, demonstrate both the beauty of mathematics (with opportunities to discover patterns and solve complex problems) and the utility of mathematics and of computational tools and methods in science, technology, and engineering. Accomplishing this will require teachers who understand mathematics well and have instructional materials able to accomplish these goals.

Similarly, an engaging and effective science education goes well beyond the low-level factual recall that is emphasized in many science classes. It must develop the skills that students need to solve complex problems, work in teams, make and recognize evidence-based arguments, and interpret and communicate complex information. These studies also support the importance of ensuring that curricula are not so crowded with topics as to drive out conceptual thinking. A clear consensus in the STEM community now surrounds these goals for science education, which can form the basis for policy decisions. In addition, many of the same principles are applicable to education in technology and engineering subject areas.

3. Political Progress. One of the most important developments of recent years has been the emergence of a clear bipartisan consensus on the need for education reform in general and the importance of STEM education in particular. In 2002 the reauthorization of the Elementary and Secondary Education Act (ESEA), renamed the No Child Left Behind (NCLB) Act, established the importance of: (1) collecting data annually about students' and schools' progress in mathematics and reading and (2) tying Federal education funding in some measure to progress. Passed by a large bipartisan majority in both the House and Senate, NCLB helped to shed light on the achievement gap and establish the principles of annual assessment and accountability. Most observers now agree that changes should be made to improve the legislation. Some have expressed concerns that the law had unintended consequences, including creating incentives for states to adopt low standards in order to ensure Federal funding, to emphasize drills at the expense of understanding, and to focus so heavily on math and reading that other subjects, including science, have received less attention. Congress is currently working on reauthorization of this important law, with modifications to improve it.

The Obama administration has sought to build on past achievements while spurring additional progress. The American Recovery and Reinvestment Act of 2009 established four broad "assurances" to improve the K-12 education system: (1) improving teacher and principal effectiveness; (2) providing information to families, educators, and researchers to improve students' schools and learning; (3) implementing college- and career-ready standards and developing improved assessments aligned with those standards; and (4) improving student learning and achievement in America's lowest-performing schools through intensive and effective interventions. The administration's *Blueprint for Reform* lays out a plan to realize these assurances through reauthorization of the ESEA and by re-envisioning the Federal role in K-12 education.[31] The administration has worked to fulfill these assurances through competitive grant-making.

An important and encouraging advance has been the work since 2008 of a state-led initiative to forge clear, consistent, and higher standards for mathematics and English language arts education in grades K-12 that can be shared across states.[32] This effort is being led by the National Governors Association and the Council of Chief State School Officers, with support and input from a variety of educational and business organizations. Common mathematics standards were released in June 2010, reflecting an unprecedented degree of cooperation on K-12 mathematics education. As of this report's publication in September 2010, 36 states and the District of Columbia had adopted the common core standards in mathematics and English language arts. There is also considerable interest in the adoption of similar standards for science. In July 2010, the National Research Council released for public comment its draft framework to guide the development of new standards for K-12 science education, and a final version will be published in early 2011. The organization Achieve, Inc., will work with a group of states to develop standards for K-12 science aligned to this framework by the end of 2012.[33]

While it is still uncertain whether the new science standards will be incorporated into existing common standards efforts, the implementation of common standards in both mathematics and science will be tremendously important. While elementary and secondary education in the United States is primarily a state responsibility, standards in mathematics, science, technology, and engineering that are widely shared among the states would serve both the states and the Nation. They would allow cross-state comparisons to guide improvements in education systems. They would drive more focused, coherent programs to prepare and support teachers. They would create larger markets for new and more effective instructional materials and technologies, along with high-quality assessments that measure all the important aspects of science learning. And they would reflect the reality that U.S. students will compete for jobs and work in a national and international economy driven by advances in technology.

Purpose of this Report

Within the overall context of education reform in the Obama administration, STEM education plays an essential role. President Obama underscored the importance of STEM education in his speech to the National Academy of Sciences in April 2009, saying that the United States needs to move its students "from the middle to the top of the pack in science and math over the next decade." He also asked the STEM community to use "love and knowledge of science to spark the same sense of wonder and excitement in a new generation." In launching the Educate to Innovate campaign in November 2009, he spoke of "strengthening America's role as the world's engine of scientific discovery and technological innovation" and declared "the improvement of STEM education over the next decade a national priority."

In the fall of 2009, the President asked his President's Council of Advisors on Science and Technology (PCAST) to develop specific recommendations concerning the most important actions that the administration should take to ensure that the United States is a leader in STEM education in the coming decades In responding to this charge, PCAST decided to focus initially on the K-12 level. A subsequent report will address STEM education at community colleges, four-year colleges, and universities. (Box 1-1 defines the scope of STEM education at the K-12 level, as used in this report.)

At its October 2009 meeting, PCAST met with Secretary of Education Arne Duncan, who described the administration's initiatives and future plans in education. PCAST also heard presentations on STEM education from representatives of the Office of Management and Budget, the Department of Defense, the National Science Foundation, the National Aeronautics and Space Administration, the Department of Energy, and the National Institutes of Health. In addition, speakers representing higher education, the private sector, state governments, and nonprofit organizations presented additional perspective on STEM education. PCAST's Subcommittee on STEM Education, chaired by Eric Lander and S. James Gates, then convened a preliminary meeting of leaders and innovators in education in December 2009 to gather information and help formulate key questions. In January 2010, PCAST appointed a Working Group on STEM Education containing, in addition to the members of the PCAST subcommittee, experts in curriculum development and implementation, teacher preparation and professional development, effective teaching, out-of-school activities, educational technology, and school administration. The group met weekly by teleconference and held a two-day meeting in Washington, DC, in April 2010. The Working Group received valuable input from Federal, state, and local officials, STEM education experts, STEM practitioners, publishers, and other experts. In addition, PCAST worked with the Office of Management and Budget and the Science and Technology Policy Institute to analyze Federal programs in STEM education. Drawing on all of this input, PCAST prepared this report beginning in May 2010.

There have been a number of important reports related to STEM education over the past two decades, including landmark reports that have called attention to the problem, reviews of the research literature, and recommendations concerning principles and priorities (see Box 1-2). Our goal is not to redo the work of these excellent reports – indeed, we have relied heavily on their research and findings.

The purpose of this PCAST report is instead to translate these ideas into a coherent program of Federal action to support STEM education in the United States. Here, we propose to the President and his administration specific goals and actionable recommendations within the context of current Federal efforts. We also have attempted to provide reasonable cost estimates for recommendations that may involve significant funding, and to determine whether the necessary actions can be supported within existing programs or may require new funding.

Many of the recommendations in this report can be carried out with existing Federal funding. Some of the recommendations could be funded in part through existing programs, although new authorities may be required in certain cases. Depending on these choices, the new funding required to fully fund the recommendations could reach up to approximately $1 billion per year. This would correspond to the equivalent of roughly $20 per K-12 public school student; or 2 percent of the total Federal spending of approximately $47 billion on K-12 education; or 0.17 percent of the Nation's total spending of approximately $593 billion on K-12 education.[34] Not all of this funding must come from the Federal budget. We believe that some of the funding can come from private foundations and corporations, as well as from states and districts.

Box 1-1. What is STEM Education?

"STEM education," as used in this report, includes the subjects of mathematics, biology, chemistry, and physics, which have traditionally formed the core requirements of many state curricula at the K-12 level. In addition, the report includes other critical subjects, such as computer science, engineering, environmental science and geology, with whose fundamental concepts K-12 students should be familiar. The report does not include the social and behavioral sciences, such as economics, anthropology, and sociology; while appropriately considered STEM fields at the undergraduate and graduate levels, they involve very different issues at the K-12 level.

We note that this report does not seek to define specific core or elective curricula, which are the subject of other important efforts. Our focus is on system-wide approaches for improving K-12 STEM education.

Box 1-2. Previous Studies

The following reports provide a broad range of analysis on the problems and potential of STEM education. For a more complete list, see www.stemedcoalition.org /content/reports/default.aspx.

Preparing Teachers: Building Evidence for Sound Policy. 2010. National Research Council.

Transforming American Education: Learning Powered by Technology. (The National Educational Technology Plan.) 2010. Office of Educational Technology, U.S. Department of Education.

Engineering in K-12 Education. 2010. National Academy of Engineering.

The Economic Impact of the Achievement Gap in America's Schools. 2009. McKinsey & Company.

The Opportunity Equation: Transforming Mathematics and Science Education for Citizenship and the Global Economy. 2009. The Carnegie Corporation of New York and the Institute for Advanced Study.

Learning Science in Informal Environments: People, Places and Pursuits. 2009. National Research Council.

Benchmarking for Success. 2008. National Governors Association, Council of Chief State School Officers, Achieve.

Fostering Learning in a Networked World. 2008. National Science Foundation Task Force on Cyberlearning.

Foundations for Success. 2008. National Mathematics Advisory Panel.

Out of Many, One. 2008. Achieve, Inc.

Building a STEM Agenda. 2007. National Governors Association.

Rigor at Risk. 2007. ACT.

Report of the Academic Competitiveness Council. 2007. U.S. Department of Education.

Taking Science to School. 2007. National Research Council.

A Model Curriculum for K-12 Computer Science. 2006. Association for Computing Machinery.

Tough Choices or Tough Times. 2006. National Center on Education and the Economy.

Rising Above the Gathering Storm. 2005. National Academy of Sciences, National Academy of Engineering, and the Institute of Medicine.

The New Educational Imperative: Improving High School Computer Science. 2005. Computer Science Teachers Association, Association for Computing Machinery.

Engaging Schools: Fostering High School Students' Motivation to Learn. 2004. National Research Council.

Adding It Up. 2001. National Research Council.

Before It's Too Late. 2000. The National Commission on Mathematics and Science Teaching for the 21st Century. (The Glenn Commission.)

Being Fluent with Information Technology. 1999. National Research Council.

A Nation at Risk: The Imperative for Educational Reform. 1983. The National Commission on Excellence in Education report to the Department of Education.

Structure of Report and Key Recommendations

In this report, we begin by examining the national goals and necessary strategies for successful STEM education (Chapter 2). We then consider the actions that the Federal Government should take with respect to leadership and coordination (Chapter 3), standards and assessments (Chapter 4), teachers (Chapter 5), technology (Chapter 6), students (Chapter 7), and schools (Chapter 8). Throughout the report, we make concrete recommendations.

Priorities

In addition to providing the full set of recommendations, we have sought to identify the most critical priorities for rapid action. We have therefore called out (in Box 1-3) our main conclusion (about the need for both preparation and inspiration) and our seven highest priority recommendations, which concern:

(1) Support for state-led shared standards and assessments
(2) Recruiting and training of great STEM teachers
(3) Recognition and reward of great STEM teachers
(4) Using technology to propel innovation
(5) Creating programs that foster inspiration through out-of-class activities
(6) Creating new STEM-focused schools
(7) Ensuring strong and strategic National leadership

Broadly speaking, our recommendations all fall under three main themes:

- Improved STEM teaching throughout K-12.
- Preparing and inspiring all students in STEM through learning opportunities inside and beyond the classroom.
- Deep and sustained commitment to innovation and data-driven decision making in K-12 education.

While PCAST's recommendations are directed toward the Federal Government, achieving the Nation's goals for STEM education in K-12 will require partnerships with state and local government and with the private and philanthropic sectors. The Federal Government must actively engage with each of these partners, who must in turn fulfill their own distinctive roles and responsibilities. In this context, we are encouraged by the state-led collaborative efforts and by the creation of private groups, such as the recently formed coalition, Change the Equation.

Box 1-3. Key Recommendations

In this box, we summarize our two main conclusions and our seven highest priority recommendations. All of these recommendations are directed at the Federal Government, and, in particular, we focus our attention on actions to be taken by the Department of Education and the National Science Foundation as the lead Federal agencies for STEM education initiatives in K-12.

Conclusions

To Improve STEM Education, We Must Focus on Both Preparation and Inspiration

To meet our needs for a STEM-capable citizenry, a STEM-proficient workforce, and future STEM experts, the Nation must focus on two complementary goals: We must prepare all students, including girls and minorities who are underrepresented in these fields, to be proficient in STEM subjects. And we must inspire all students to learn STEM and, in the process, motivate many of them to pursue STEM careers.

The Federal Government Has Historically Lacked a Coherent Strategy and Sufficient Leadership Capacity for K-12 Stem Education

Over the past few decades, a diversity of Federal projects and approaches to K-12 STEM education across multiple agencies appears to have emerged largely without a coherent vision and without careful oversight of goals and outcomes. In addition, relatively little Federal funding has historically been targeted toward catalytic efforts with the potential to transform STEM education, too little attention has been paid to replication and scale-up to disseminate proven programs widely, and too little capacity at key agencies has been devoted to strategy and coordination.

RECOMMENDATIONS

(1) Standards: Support the Current State-Led Movement for Shared Standards in Math and Science

The Federal Government should vigorously support the state-led effort to develop common standards in STEM subjects, by providing financial and technical support to states for (i) rigorous, high-quality professional development aligned with shared standards and (ii) the development, evaluation, administration, and ongoing improvement of assessments aligned to those standards.

The standards and assessments should reflect the mix of factual knowledge, conceptual understanding, procedural skills, and habits of thought described in recent studies by the National Research Council.

(2) Teachers: Recruit and Train 100,000 Great STEM Teachers over the Next Decade Who are Able to Prepare and Inspire Students

The most important factor in ensuring excellence is great STEM teachers, with both deep content knowledge in STEM subjects and mastery of the pedagogical skills required to teach these subjects well.

The Federal Government should set a goal of ensuring over the next decade the recruitment, preparation, and induction support of at least 100,000 new STEM middle- and high school teachers who have strong majors in STEM fields and strong content-specific pedagogical preparation, by providing vigorous support for programs designed to produce such teachers.

(3) Teachers: Recognize and Reward the Top 5 Percent of the Nation's STEM Teachers, by Creating a Stem Master Teachers Corps

Attracting and retaining great STEM teachers requires recognizing and rewarding excellence.

The Federal Government should support the creation of a national STEM Master Teachers Corps that recognizes, rewards, and engages the best STEM teachers in the nation and elevates the status of the profession. It should recognize the top 5 percent of all STEM teachers in the Nation, and Corps members should receive significant salary supplements as well as funds to support activities in their schools and districts.

(4) Educational Technology: Use Technology to Drive Innovation, by Creating an Advanced Research Projects Agency for Education

Information and computation technology can be a powerful driving force for innovation in education, by improving the quality of instructional materials available to teachers and students, aiding in the development of high-quality assessments that capture student learning, and accelerating the collection and use of data to provide rich feedback to students, teachers, and schools. Moreover, technology has been advancing rapidly to the point that it can soon play a transformative role in education.

Realizing the benefits of technology for K-12 education, however, will require active investments in research and development to create broadly useful technology platforms and well designed and validated examples of comprehensive, integrated "deeply digital" instructional materials.

The Federal Government should create a mission-driven, advanced research projects agency for education (ARPA-ED) housed either in the Department of Education, in the National Science Foundation, or as a joint entity. It should have a mission-driven culture, visionary leadership, and draw on the strengths of both agencies. ARPA-ED should propel and support (i) the development of innovative technologies and technology platforms for learning, teaching, and assessment across all subjects and ages and (ii) the development of effective, integrated, whole-course materials for STEM education.

(5) Students: Create Opportunities for Inspiration through Individual and Group Experiences Outside the Classroom

STEM education is most successful when students develop personal connections with the ideas and excitement of STEM fields. This can occur not only in the classroom but also through individualized and group experiences outside the classroom and through advanced courses.

The Federal Government should develop a coordinated initiative, which we call INSPIRE, to support the development of a wide range of high-quality STEM-based after-school and extended day activities (such as STEM contests, fabrication laboratories, summer and afterschool programs, and similar activities). The program should span disparate efforts of science mission agencies and after-school programs supported through the Department of Education funding.

(6) Schools: Create 1,000 New STEM-Focused Schools over the Next Decade

STEM-focused schools represent a unique National resource, both through their direct impact on students and as laboratories for experimenting with innovative approaches. The Nation currently has only about 100 STEM-focused schools, concentrated at the high school level.

The Federal Government should promote the creation of at least 200 new highly-STEM-focused high schools and 800 STEM-focused elementary and middle schools over the next decade, including many serving minority and high-poverty communities. In addition, the Federal Government should take steps to ensure that all schools and school systems have access to relevant STEM-expertise.

(7) Ensure Strong and Strategic National Leadership

Stronger leadership, coherent strategy and greater coordination are essential to support innovation in K-12 STEM education. Toward this end, the Federal Government should (i) create new mechanisms, with substantially increased capacity, to provide leadership within each of the Department of Education and the National Science Foundation; (ii) establish a high-level partnership between these agencies; (iii) establish a standing Committee on STEM Education within the National Science and Technology Council responsible for creating a Federal STEM education strategy; and (iv) establish an independent Presidential Commission on STEM Education, in conjunction with the National Governors Association, to promote and monitor progress toward improving STEM education.

PCAST believes that the Nation has an urgent need – but also, thanks to recent developments, an unprecedented opportunity – to bring together stakeholders at all levels to transform STEM education to lay the groundwork for a new century of American progress and prosperity.

Because we consider STEM education a critical national priority, PCAST plans to maintain a sustained focus on the topic, recognizing that the problems and opportunities identified in this report require an ongoing effort over the long term. PCAST looks forward to continued consultations on a regular basis with the Department of Education and National Science Foundation, ideally beginning in about six months, to discuss progress and challenges in achieving the Nation's goals for STEM education.

II. Preparation and Inspiration

Chapter Summary

The Nation requires clear goals and a coherent strategy for improving K-12 STEM education. This report attempts to articulate the key priorities and to define an approach that will guide how STEM education initiatives unfold. Increasingly, all citizens will need to be able to use scientific knowledge, advanced technologies, and quantitative methods in their jobs and to make informed decisions about critical issues. The United States will also need a steady stream of the world's best researchers and innovators to remain at the leading edge of science and technology. In addition, all citizens should have the opportunity to experience the wonders of discovery and exploration available in STEM. To meet the Nation's needs for a STEM-capable citizenry, a STEM-proficient workforce, and future STEM experts, we recommend a strategy that focuses on two complementary goals: We must *prepare* all students, including girls and minorities who are underrepresented in these fields, to be proficient in STEM subjects. And we must *inspire* all students to learn STEM and, in the process, motivate many of them to pursue STEM careers.

Introduction

Improving the quality and effectiveness of STEM education in the United States requires a clear understanding of both goals and strategies. In this chapter we outline the national need in terms of four goals that need to be achieved to have a well-educated citizenry and workforce. We also discuss the distinctive aspects of STEM education that require unique approaches within the context of education reform. Finally, we discuss the two components that must underlie our national strategy: We must **prepare** students so they all have a strong foundation in STEM and are able to pursue it. And we must **inspire** students so that all are motivated to learn STEM and many are excited about entering STEM fields.

National Needs for STEM Education

Given the critical role that science and technology will play in our Nation's future, four key areas of national need are apparent:

1. **We must ensure a STEM-capable citizenry**. All U.S. citizens should have an understanding of scientific and technological knowledge, engineering principles, and quantitative methods sufficient to succeed in public life and in their careers, and to make informed decisions about issues facing our Nation and our planet. Learning STEM is important even for people who will never become engineers, mathematicians, or scientists. An increasing number of jobs draw on at least some knowledge and skills from STEM fields – such as data analysis, problem solving, and the ability to analyze and use evidence – and every occupation has the potential to be transformed by scientific and technological advances. In addition, when American citizens make personal decisions, engage in civic discourse, serve on a jury, cast their ballots, or run for public office, they should have the knowledge, conceptual understandings, and critical-thinking skills that come from studying STEM subjects.
2. **We must build a STEM-proficient workforce**. The U.S. economy needs a large and increasing supply of workers who can routinely use scientific, technological, engineering, and mathematical knowledge and skills in their jobs; this knowledge fuels innovation and entrepreneurship. The nation's ability to solve problems and propel economic growth will therefore depend on cultivating a future workforce that is *STEM-proficient*. The STEM-proficient workforce currently comprises as many as 21.4 million people – approximately 15 percent of the employed population.[35] Furthermore, employment in STEM fields is increasing at a faster pace than in non-STEM fields.[36] Even during the recent recession, companies in STEM-related fields, such as in the aerospace, defense, life sciences, and energy sectors, reported

shortages of skilled workers, and these shortages are expected to persist.[37] Moreover, many professions once perceived as not requiring STEM skills, such as agriculture and law, increasingly require technological and scientific proficiency. The Business and Industry STEM Coalition has called for the country to double – to 400,000 – the number of college graduates with degrees in STEM per year by 2020.[38] Our education system must prepare more people who can work with, build, and adapt to technology to expand opportunities for all Americans and to ensure that the Nation's living standards continue to rise. Preparing this workforce is essential to the country's future prosperity.

3. **We must cultivate future STEM experts.** The United States also needs a steady stream of the best STEM researchers and innovators in the world. These individuals will come up with new ideas and inventions, pioneer new fields and industries, and inspire and mentor new generations of scientists, engineers, and mathematicians. They will make the fundamental discoveries that help us understand our world. They will keep the United States at the leading edge of science and technology in the 21st century. Scientists, engineers, and mathematicians have historically founded and made tremendous contributions to high- tech industries, medical research centers, engineering firms, and government agencies in the United States. They have contributed immensely to economic growth, to technological progress, to our understanding of ourselves and the universe, and to the reduction of hunger, disease, and poverty. STEM experts also provide a wealth of knowledge that serves national security and protects our natural resources, and they attract students and investors from around the world to the United States. If we hope to retain these advantages, even as other countries seek to build their STEM expertise, we must cultivate a large pool of STEM experts with the knowledge, drive, and imagination to advance the frontiers of science and industry.
4. **We must close the achievement and participation gap.** Our national needs cannot be met without drawing on the full potential of our Nation. The United States cannot remain at the forefront of science and technology if the majority of its students – in particular, women and minorities underrepresented in STEM fields – view science and technology as uninteresting, too difficult, or closed off to them. We must close the achievement and interest gap in STEM subjects among racial, ethnic, and gender groups. Closing these gaps cannot be limited to helping students and groups at the remedial level in STEM subjects. It also requires unleashing the full potential of all our students who have not historically been drawn to STEM fields. STEM education needs to recognize and cultivate untapped talent. Many of our future STEM experts can and must come from traditionally underserved populations. STEM fields will greatly benefit from drawing on a diversity of perspectives, cultures, and ideas.

Given these goals, STEM education policies must be aimed at multiple levels and at everyone. We must ensure that struggling students reach STEM-proficiency. In parallel, we must deeply engage proficient students and attract high-achieving students from all groups to STEM subjects.

Distinctive Nature of STEM Education

Efforts to improve K-12 STEM education must fit within the broader context of K-12 education reform within the United States. We will not be able to transform STEM education unless we solve systemic issues that prevent our schools and school systems from fulfilling their potential. One fundamental issue is a failure of accountability: we fail at many levels to define clear goals, measure our success toward these goals and hold ourselves accountable for our performance. A second fundamental issue is a failure to reward excellence: we know that teachers are the most critical component in the educational system, yet we fail to instill, recognize, and reward excellence in teachers. Given these two problems, it is no surprise that many schools are in trouble.

The administration has taken a comprehensive approach that attempts to address these issues at every level of the education system. Of particular importance, Secretary of Education Arne Duncan has articulated four "assurances" that should form the foundation for reform efforts. These are: (i) adopting rigorous standards that prepare students for success in college and the workforce; (ii) recruiting and retaining effective teachers, especially in classrooms where they are needed most; (iii) turning around low-performing schools; and (iv) building data systems to track student achievement and teacher effectiveness. We endorse these principles and agree that they provide a foundation for general education reform.

At the same time, it is important to recognize that STEM education has a number of distinctive characteristics that pose both unique challenges and opportunities. Achieving the President's goal of dramatically improving STEM education will require the development of distinctive approaches within the larger overall framework. These considerations include:

- **STEM subjects tend to be highly cumulative and sequential.** In mathematics, each step in a progression depends on previous knowledge and skills. If students fail to understand ratios and fractions or properties of matter, for example, they are likely to fall further behind in the mathematics or science courses that follow. Science courses, for their part, are enhanced by cross- disciplinary knowledge that transcends the typical course boundaries of biology, chemistry, earth science, or physics. These characteristics of STEM mean that students who have trouble at an early stage will face further difficulty down the road; it is easy to get off the path and hard to get back on. It also means that teachers need to have knowledge that goes beyond their specific course and the confidence to use it to help guide and enhance student understanding and achievement.
- **STEM knowledge is specialized.** Some of the knowledge and methods of STEM subjects can be difficult for students to master in the context of their everyday lives. Teachers at all grade levels need deep content knowledge to be able to explain basic concepts well, as well as to answer deeper questions from inquisitive students. They must also be able to anticipate and correct naïve notions that learners of mathematics, science, and technology bring to the classroom. They must be able to teach those subjects in different ways to reach different students.
- **STEM knowledge is rapidly changing.** The frontiers of knowledge in STEM fields are ever- expanding, as scientific progress and technological advances constantly reshape our understanding of the human body, the cosmos, the complex dynamics of

our climate and the Earth's ecosystems, and the potential of technological tools. This rapid change presents unique opportunities for engaging learners by connecting them current explorations and investigations. And, as the nature of scientific practices changes to become more global, collaborative, and data- intensive, opportunities for STEM learning also expand. But taking advantage of these opportunities requires that teachers stay current with subject areas that may have changed substantially since they left college. This characteristic of STEM fields underscores the importance of conceptual understandings rather than memorization of facts that can become outdated.

- **STEM-trained individuals have alternative, high-paying career options.** Many of the highest paying professions for recent college graduates are related to STEM fields.[39] This means that the most proficient STEM students in colleges may be attracted to careers other than teaching. Attracting and retaining the best-trained STEM students to the teaching profession will require unique strategies that may differ from those required to recruit teachers in other fields.
- **STEM is not always familiar and accessible to the public and education leaders.** Many parents, school administrators, state and district officials, and public leaders have only limited familiarity, engagement, or mastery of STEM subjects. Although opinion polls show that most of the public believes science has a positive and significant impact on society, [40] the general public's comfort with STEM subjects is much more limited. Many elementary school teachers, because of their preparation, tend to view STEM subjects with more trepidation than other subjects, and may shy away from teaching them. This anxiety is often also reflected in schools and colleges of education that prepare teachers. STEM teachers at upper grade levels often bemoan the lack of support for STEM on the part of their schools and school district leaders. At a time when the need for more support for STEM education is critical, many state and district STEM teams are being down-sized or eliminated. In addition, many school principals are unfamiliar with the knowledge and strategies needed to support and recognize high-quality STEM classrooms in their schools. This makes it difficult to gain traction for innovative ideas, curricula, and resources for STEM education.
- **The STEM professional community can be a major resource for educators.** The United States has a highly developed and accomplished STEM community that can be engaged effectively in the important work of teaching and mentoring students. Scientists, mathematicians, engineers, and technologists can be found throughout academia, industry, government, and nongovernmental organizations. Many STEM professionals would be willing to contribute their knowledge, time, and skills to educating youth if effective ways could be found to engage them in a way that is useful to teachers and students. Ideally, teachers in schools and out would be able to tap into the expertise and rapidly advancing knowledge in universities, corporate and government laboratories, offices, and field stations to expand learning opportunities for their students.
- **STEM-trained individuals tend to be tech-savvy.** STEM can be an ideal testing ground for the use of technology and innovative learning tools in education. STEM practitioners tend to be most able to create such tools, and STEM teachers may be particularly comfortable with technology-based innovation.

The distinctive characteristics of STEM education and the STEM community merit special consideration in formulating U.S. education policy. They should shape the approaches taken to preparing teachers, creating instructional materials, and enhancing schools for K-12 STEM education. The distinctive attributes of STEM also hold great potential. They provide opportunities to engage and educate students in real- world settings and connect with the career paths that will be available to them in an increasingly technology-intensive economy.

Strategy: Prepare and Inspire

There are two obstacles to achieving the nation's goals with respect to STEM education. First, too few U.S. students are proficient in STEM. Second, too few of those who are proficient pursue STEM fields. Of all ninth graders in the United States in 2001, for example, only about 4 percent are predicted to earn college degrees in STEM fields by 2011,[41] a clear indicator that at various stages of the education system we are losing potential STEM talent.

This loss begins well before high school. In both mathematics and science, about 70 percent of eighth graders score below the proficient level.[42] Students who lack proficiency face a mounting barrier, as it becomes increasingly difficult to engage in STEM subjects without a solid foundation in basic skills, such as algebra. Even among the minority of students who are proficient in STEM in eighth grade, about 60 percent decide during high school that they are not interested in these subjects.[43] Of those who remain interested in high school, only about 40 percent actually enter STEM majors in colleges. Of these students, 60 percent switch out of STEM[44] while far fewer switch in. The problem is particularly acute among minority groups and women. Black, Latino, and Native American students who show interest in STEM as college freshmen are much less likely to graduate with STEM degrees than their white and Asian American counterparts.[45]

To address these challenges, we need a two-pronged approach: (1) we must *prepare* all students so they have a strong foundation in STEM no matter what careers they pursue, and (2) we must *inspire* students so that all are motivated to learn STEM subjects and many are excited about entering STEM fields. Preparation involves building shared skills and knowledge. Inspiration involves individual, meaningful experiences that speak to students' particular interests and abilities.

Preparation involves bringing all students up to the level of proficiency in STEM subjects. This requires a focus on high standards and meaningful assessments for STEM subjects, together with providing teachers and schools with tools to enable active, engaged learning by students. It also requires recruiting, preparing, and retaining teachers who have the knowledge and skills to effectively convey those standards. And it requires providing schools and teachers with tools and support to help students achieve these standards. Preparation includes giving the nation's high-achieving STEM students challenges and opportunities to reach even higher. This requires recognizing and nurturing talent in all groups of students. Preparing students to succeed and excel in STEM, as well as in subjects and careers that draw upon STEM aptitudes and knowledge, will require a substantial effort that addresses the current inadequacies at all levels of the education system.

Inspiration involves capturing the curiosity and imagination of students. If preparation depends heavily on skill development, inspiration depends on providing access to exciting individual experiences and to STEM connections inside and outside of schools. Inspiration also involves giving students the opportunity to be motivated by teachers and mentors, by collaborations in discovery and invention, and by what they learn in school and out of school.

Students need exciting experiences that speak to their interests – in school among teachers, peers, and mentors, beyond the curriculum, and beyond the classroom. These experiences should reveal to them the satisfaction of solving a problem, discovering a pattern or phenomenon on one's own, becoming insatiably curious about a puzzling question, or designing and creating an invention. Students should be able to see themselves in the role of a scientist, technologist, engineer, or mathematician, which often requires an association with role models. They need to have experiences that demonstrate STEM subjects can connect to their own lives and how working in these fields can help solve some of the most serious problems human beings have ever faced. Research shows how important it is for children to have exciting experiences in STEM early on, in elementary and middle school, to capture their interest and spark a lifelong passion. Students who express interest in STEM in eighth grade are up to three times more likely to ultimately pursue STEM degrees later in life than students who do not express such an interest.[46, 47]

Inspiration can come from many sources. When setting the Nation's goals for STEM education, we must take advantage of the opportunity to inspire students through the many avenues opened up by technological advances and promising programs in a variety of settings. We will have more to say about the need to inspire students throughout this report.

Implications

These complementary objectives – preparation and inspiration – can be the driving forces behind an overall strategy in STEM education. We need states to adopt high standards, well-designed assessments, and reliable accountability measures. We also need truly excellent teachers who are deeply knowledgeable about STEM content and pedagogy. We need materials that are challenging, stimulating, and up to date with scientific and technological knowledge. We need learning opportunities both inside and outside the classroom for students to explore, invent, and discover. We need all teachers and schools to have connections to the dynamic, broader world of STEM. We need school and school district officials who can develop and lead schools that enable students to learn to their fullest potential. In subsequent chapters, we describe actions that the Federal Government can take to help achieve these objectives.

III. Federal Role in K-12 STEM Education

Chapter Summary

Although K-12 education is largely the responsibility of the states and localities, Federal funding can have a profound effect on STEM education by leveraging state and local resources in the pursuit of national goals. For this potential to be realized, Federal involvement needs to reflect a coherent and coordinated strategy. Today the Federal Government spends around $1 billion[48] on K-12 STEM education, and the Obama administration has emphasized STEM education in several major policy initiatives. Nevertheless, these efforts represent far too small a proportion of the overall investment in K-12 education, and Federal activities should be more strategic. This requires strong leadership and new policy and funding mechanisms in the National Science Foundation, the Department of Education, and the Executive Office of the President. The Federal Government also should draw on expertise from business, academia, and other levels of government to foster excellence in K-12 STEM education.

Introduction

Elementary and secondary education is largely a state and local responsibility in the United States, with the Federal Government typically contributing only about 8 percent[49] of education funding. Nonetheless, this Federal funding can play a crucial role in improving U.S. education if it is used wisely to advance national priorities.

Federal support for K-12 STEM education flows through many different Federal agencies. The Department of Education funds policies and programs that have a substantial influence on schools and students across the Nation. Funding also flows through science mission agencies such as the National Science Foundation, which supports activities to develop instructional materials and educational technologies as well as activities that affect a subset of teachers and students.

For this Federal funding to constitute an investment in the Nation's future worth more than the dollars disbursed, it must reflect a coherent strategy for STEM education. Coherence in turn requires widely shared goals as well as structures that ensure leadership, coordination, and oversight. Despite clear recognition of the importance of STEM education, Federal policy over the past quarter-century has lacked such coherence and leadership. In this chapter, we describe Federal funding for K-12 STEM education and offer recommendations to improve the coherence, leadership, and coordination of Federal programs.

Funding for K-12 STEM Education

The K-12 education system in the United States is a large and complex enterprise. There are nearly 50 million public school students and 3.2 million teachers in 98,000 schools organized into 14,000 school districts. (These figures do not include K-12 private school students and teachers.) There are approximately 1.7 million elementary school teachers, very few of whom are specialists in STEM subjects, and about 1.3 million middle and high school teachers, of whom about 425,000 have a main teaching assignment in mathematics or science.[50, 51] Total expenditures for public elementary and secondary education are now about $593 billion per year.[52] Historically, Federal funding has contributed about 8 percent of this total, with states and localities contributing roughly 47 percent and 44 percent, respectively.[53]

In the last two decades, Federal spending in education has been used to advance certain policy objectives. In 1994, two major laws affecting STEM education were enacted: the Goals 2000: Educate America Act,[54] which placed heightened emphasis on mathematics and science education, especially in the early grades, and on increasing teachers' substantive background in mathematics and science; and the Improving America's Schools Act, which sought to help disadvantaged students meet high standards in academic disciplines, including mathematics and science, and initiated a trend of increased state assessments.[55] The Elementary and Secondary Education Act (also known as the No Child Left Behind Act), which President Bush signed into law in 2002, focused attention on school accountability and demonstrated how profoundly strategic changes in Federal policy can drive changes in schools across the country. In addition, the America COMPETES Act of 2007 authorized a range of STEM education programs in the Department of Education and the science mission agencies.[56]

Below, we describe briefly the current landscape of Federal expenditures on K-12 STEM education in the Department of Education and the Federal science mission agencies in FY10 and via the American Recovery and Reinvestment Act (ARRA). We also describe some of the proposals made in the President's budget for FY11. Our analysis is based on an inventory of Federal programs assembled by the Office of Management and Budget. We then discuss methods and mechanisms to evaluate the effectiveness of these programs.

Funding for STEM Education at the Department of Education

Most of the Federal funding for elementary and secondary education, approximately $25 billion in discretionary funding in FY10, flows through the Department of Education under appropriations authorized under various titles in the Elementary and Secondary Education Act (ESEA).[57] Funding under ESEA flows through a series of titles.[58]

We describe below those funds that are targeted specifically to STEM education. It is important to note that Titles I and II include substantial funds that the states themselves may choose to use for mathematics or science education. Our focus here, however, is to identify funds that the Federal Government itself targets to STEM activities, since it is important to ensure that they are spent strategically.

- ***Title I ($15.9 billion).*** Under NCLB, more than 50,000 public schools across the country receive funds to provide additional academic support and learning opportunities to help low-achieving children master challenging curricula and meet state standards in core academic subjects. For example, these funds support extra instruction in reading and mathematics as well as special preschool, after-school, and summer programs to extend and reinforce the regular school curriculum.[59] Much of this funding is awarded to states and local education agencies by formulas linked to the number of students from low socioeconomic backgrounds. About $45 million is specifically appropriated to increase the number of students who take Advanced Placement (AP) courses and exams, including AP courses and exams in mathematics and science subjects.[60]

- ***Title II ($3.2 billion).*** Funds under Title II target the preparation and recruitment of teachers and principals ($2.9 billion). A small portion ($180 million in FY10) is targeted to mathematics and science education through formula-driven Mathematics and Science Partnerships, which are intended to improve the content knowledge of teachers and the performance of students in the areas of mathematics and science by encouraging collaborative programs among states, higher education institutions, local education agencies, and elementary and secondary schools. In addition, $100 million is targeted to educational technology[61] through state grants. In general, states apply to the Department of Education for these grants, and applications must include plans for how states will allocate funds to local education agencies (and ensure compliance), how activities will be tied to state standards, and how grants will improve the quality of teachers and principals.
- ***Title IV.*** Among the funds allocated under Title IV are $1.16 billion to support after-school programs through the 21st Century Community Learning Centers (CCLC). These centers, which receive funding through a competitive grants program, provide opportunities for academic enrichment, particularly for students who attend low-performing schools, to help them meet state and local achievement standards in core academic subjects such as reading and mathematics. These programs may include STEM-related activities; for example, 25 states reported using a combined total of $156 million on grants for science and technology activities in 2008–09.[62]

Another source of support for STEM education within the Department of Education is the Institute of Education Sciences (IES), which was funded at a level of $659 million in FY10. The Institute, which was created in 2002, funds research, regional education laboratories, assessment (including the National Assessment of Educational Progress and the National Assessment Governing Board), and statewide data systems. According to data supplied by IES, approximately $45 million is related in some way to research on STEM education.[63]

The Department also provides about $96 million in Federal student aid (mostly in the form of loans and loan forgiveness programs) to prospective science and mathematics teachers who agree to teach in needy schools.[64]

In total, the Department of Education funding includes approximately $522 million per year that is targeted exclusively for STEM education. We focus here on funds targeted specifically to STEM education, because they are the mechanisms by which Federal policy can directly guide improvements in STEM education. As noted above, the states also receive significant formula grants under Titles I and II (focused on providing additional support for students living in poverty and on increasing teacher and principal quality) that they may use at their discretion, including on STEM subjects. Thus, the total amount of Federal funding actually used for STEM education is larger than the amounts discussed here.

The Obama administration has advanced several initiatives to increase the Department of Education's commitment to STEM education:[65]

- The American Recovery and Reinvestment Act of 2009 included several initiatives that have the potential to influence K-12 STEM education. The Race to the Top (RTTT) incentive grants received $4.35 billion, and $650 million went for the Investing in Innovation (i-3) Fund. STEM was a competitive priority for RTTT funding, meaning that an application could receive bonus points for including a focus on STEM. In FY11, the Investing in Innovation Fund will direct $150 million of its $500 million budget toward STEM projects and will include a cross-cutting emphasis on technology. We strongly encourage the administration to continue these programs, which provide critical support for STEM initiatives.

Table 3-1. Department of Education Funding (FY09 ARRA, FY10) and Funding Requests (FY11) Specifically Related to STEM Education

Program	FY09 ARRA and FY10 Funding	Approx. Amount Designated for STEM	Program	FY11 Proposed Funding	Approx. Amount Designated for STEM
ESEA Title I	$15.9 billion	Most not designated specifically for STEM, but approximately half ($8 billion) goes to math (Title I funding split between math and English language arts), but this is not considered STEM specific in this report's analysis*	Various	$14.5 billion	Approximately half of total ($7.25 billion) would go to math, but this is not considered STEM-specific in the report's analysis*
		Includes $45 million for advanced courses, including AP and IB (not math or science specific, but includes support for those courses)	All Students College and Career Ready[1]		Proposed $100 million for "college pathways and accelerated learning"; funding is not specific for AP.[7]
ESEA Title II	$3.2 billion	$180 million for Math and Science Partnerships (MSP)	Effective Teaching and Learning for a Complete Education[2]	$1.02 billion	MSP consolidated into $300 million "effective teaching and learning: STEM" program within this larger $1.02 billion program.
			Effective Teachers and Leaders[3]	$3.85 billion	No funding specified for STEM, but activities encompass STEM.
ESEA Title IV: 21st Century Learning Communities	$1.16 billion	$156 million[7]	21st Century Learning Communities[4]	$1.16 billion	At least $156 million (if funding for STEM programs remains the same as in previous years).[8]
Institute for Education Sciences	$659 million	$45 million	Institute for Education Sciences[5]	$739 million	$55 million
Race to the Top	$4.35 billion (ARRA)	Not specified	Race to the Top	$1.35 billion	STEM may be a competitive criteria.
Investing in Innovation (i3)	$650 million (ARRA)	Not specified	Investing in Innovation[6]	$500 million	$150 million for STEM projects
Loan Forgiveness		$96 million for math and science teachers	Loan Forgiveness		$106 million
TOTAL		**$522 million**			**$867 million***

* Funding total for Department of Education includes only STEM-specific funding as reported by the Department of Education for the OMB inventory and the internal analysis of 21st Century Community Learning Centers. Totals do not include formula driven funding to states under Title I in FY 09 ARRA or FY 10 or the Administration's FY 11 budget request.

Notes: [1] To support statewide accountability systems linked to standards and assessments based on the expectation that all students will graduate from high school college- and career-ready. Schools and local education agencies that make significant progress toward the goal and close achievement gaps will be rewarded. Rigorous interventions will be carried out in the lowest performing schools.

[2] To provide competitive grants to state agencies, including agencies partnering with outside entities, and to improve teaching and learning in low-performing schools. The MSP program from FY10 is consolidated into this program, with a focus on effective teaching and learning of STEM.

[3] To target the preparation, training, and recruitment of teachers and principals. There are three priorities for FY11: $2.5 billion to the effective teachers and leaders programs to focus on the equitable distribution, design, and implementation of rigorous and fair teacher evaluation systems and the provision of high-quality professional development; $950 million to the teacher and leader innovation fund, which creates incentives for effective teachers and school leaders to work in the most challenging schools; and $405 million to Teacher and Leader Pathways, a new program to support the creation or expansion of high-quality pathways into teaching and to support new principals and school leadership teams focused on school turnaround.

[4] To emphasize extended learning opportunities and redesigning or extending the school day, week, or year to provide additional time for academic and enrichment activities.

[5] To support an increase in research, development, and dissemination; statistics; assessment; and statewide data systems.

[6] To develop and validate promising practices, strategies, and programs.

[7] The ESEA reauthorization proposal shifts Federal support for advanced courses to a program in Title II. The budget request includes $100 million for proposed "college pathways and accelerated learning", although funding is not specific for AP or for science and math courses.

[8] This level of funding for FY10 is estimated, derived from state-reported funding in for FY09 of 1,622 grants for science and technology activities in 25 states that report individual, rather than aggregated, activities for the 21st Century Community Learning Center grants. This funding does not include tutoring, supplemental educational services, or homework help. Grants are awarded to more than 15 kinds of organizations, including local education agencies, community-based organizations, and institutions of higher education, which report that about 32 percent of their activities, target STEM.

Table 3-2. Science Mission Agency Funding for K-12 STEM Education, Teachers, and Outreach (estimated)[1]

Agency	Recovery Act ($MM)	FY10 Funding ($MM)	FY11 Budget Request ($MM)
NSF	$93.0	$458.3	$453.7
NASA		$87.8	$89.5
HHS/NIH		$45.8	$43.7
Commerce (NIST, NOAA)		$40.1	$32.2
DOD		$22.9	$23.6
Energy		$11.1	$26.5
EPA		$3.4	$ 3.6
USDA		$1.5	$4.0
DOT		$1.25	$1.25
NSA		$0.4	$0
TOTAL[2]	**$93.0**	**$672.6**	**$678.1**

[1] In response to a PCAST request, OMB requested data from Federal agencies on their STEM Education programs. These data were then collected in a central database by the Science and Technology Policy Institute (STPI) and expanded with additional information on the purpose of each of the STEM programs. For this report, PCAST analyzed data in the central database.

[2] Agency total for STEM education do not include funding from the U.S. Department of Labor, which targets primarily workforce training through the Employment and Training Administration for those older than age 16. (These funds total over $5.4 billion in FY10 funding plus $4 billion in FY09 ARRA funding.) The inventory identified no funded K-12 outreach or informal education programs in the Department of Interior or Department of Homeland Security.

- Funding The President's FY11 budget proposes significant changes to increase the Department of Education's efforts in STEM education and to consolidate several programs into targeted areas. As shown in Box 3-1, these changes would reorganize programs under Titles I, II, and IV of the Elementary and Secondary Education Act as well as several other programs. These specific and strategic commitments represent a significant increase for STEM education compared with the investment in the previous decade. They signal the Department's desire to take a more active role in using its policies and funding strategically to build a sustainable infrastructure to improve STEM education in schools.

g for STEM Education at Science Mission Agencies

In addition to the Department of Education, various Federal science mission agencies provide support for STEM education. These science mission agencies collectively will spend \$3.6 billion on STEM education in FY10, but the vast majority is spent on higher education programs such as graduate and postgraduate training and fellowships. Only about \$673 million, or 19 percent of the total spent by science mission agencies, is allocated for formal, informal and outreach programs affecting K-12 STEM education and teachers, as shown in Table 3-2.[66] The FY11 budget proposes relatively minor changes in these amounts.

National Science Foundation

Among the science mission agencies, the National Science Foundation (NSF) is by far the largest funder of K-12 STEM education. The NSF's commitment to integrating research and education makes it uniquely positioned to bring cutting edge STEM content and practices to investments in education. The total amount spent on science and mathematics education is similar in magnitude to the funding from the Department of Education. The funding is administered by several programs under the Education and Human Resources Directorate. The largest programs are described in Box 3-1.

Since its inception in 1950, the NSF has played an active role in STEM education. It was created as an independent agency to "develop and encourage the pursuit of a national policy for the promotion of basic research and education in the sciences."[67] At various times in its history – especially in the 1960s through the 1990s – the NSF provided critical support to science and mathematics teachers. It also has funded bold experiments to develop curricula, instructional materials, educational technologies, and programs to engage the public. Some of the NSF-funded activities in research and evaluation have charted new directions for the field. It has a deep reservoir of information about what has been tried – and what has succeeded and failed – in science and mathematics education.

At its best, the NSF has excelled in funding investigator-driven experimental research in STEM education. This approach led to major investments in the research and early development phases of innovative full- course K-12 curricula in mathematics, biology, and physics. These investments also encompassed professional development for teachers (often during summer institutes) to learn how to use and incorporate the curricula into their classes. The large-scale curricula developed during the 1960s and 1970s involved research on what kinds of knowledge should be taught and how that knowledge should be taught. This became the basis for the development of prototypical curricular materials that then were field-tested and, in some cases, licensed to commercial publishers. Nevertheless, only a small number of these materials achieved widespread use in K-12 classrooms.

By the late 1970s, these investments in large-scale curriculum development (representing full, coherent K-12 programs of study in mathematics and the sciences) and related teacher professional development (including through the NSF institutes) largely disappeared. Most of the teachers who were beneficiaries of those Federal investments have retired. In the mid-1980s and early 1990s, investments focused more on students' understanding of STEM concepts.[68] The NSF resumed support for the development of instructional materials in the 1980s with the launch of the Instructional Materials Development (IMD) program, which lasted until it was replaced in 2006 with the current Discovery Research K-12 program (DR K-12). The IMD program resulted in investments in several comprehensive (multi-grade and full-year) mathematics and science curricula.[69]

Currently, the NSF funds proposals to develop instructional materials that focus on specific topics as well as "comprehensive" materials that are intended to cover a semester or a year or to examine the trajectory of a single concept over several years. The current DR K-12 research and development portfolio on instructional materials includes applied efficacy research – studies that look at how, for whom, why, and under what conditions innovative instructional materials work. The focus on these studies over the last four to five years signals an additional emphasis for the NSF and is intended to improve the clarity and understanding of how particularly difficult mathematics and science concepts are addressed in curricula. Findings from this work are just emerging. In the Research and Evaluation on Education in Science and Engineering (REESE) portfolio, research looks at more basic and complex problems in learning and education, often in the context of developing innovative instructional approaches.

A program known as the Teacher Preparation and Enhancement Program was directed at STEM teaching in the mid-1980s. It was followed by the Systemic Initiatives in the 1990s, which made a major investment in teacher professional development, and then by the Teacher Professional Continuum program in the early 2000s. Since 2005, teacher professional development has been funded largely through the Mathematics and Science Partnership (MSP) teacher institutes. The MSP program's comprehensive projects are aimed at states, regions, or large districts.[70] A small amount of funding has been awarded through other NSF programs for discipline-specific research experiences for teachers in academic settings. Aspiring mathematics and science teachers are funded through the Robert Noyce Fellowship program.

The NSF also funded programs in the 1990s to build state and local capacity to support strong science and mathematics instruction through its state, rural, and urban Systemic Initiatives programs. These initiatives ended in 2005 and were replaced by the MSP program, thus shifting away from a focus on developing and enabling mathematics and science education leaders at the state and local levels to implement policies and programs based on a rigorous R&D process.

The NSF has sought to make the many learning tools and instructional materials it has funded available to teachers via the National STEM Digital Library (NSDL). However, NSDL is largely a collection of disparate tools and independent modules, and its reach has been limited. While some teachers can make excellent use of such resources, many teachers require more integrated, ready-to-use instructional materials tied to their curricula and appropriate to their classrooms.

In addition to its programs devoted specifically to K-12 education, the NSF has sought to leverage its scientific grants to have an impact on STEM education. Since 1997 the NSF has evaluated all research proposals on the basis of both intellectual merit and broader societal impacts.[71] Many grantees respond to the broader impact criterion by using a portion of investigators' grants to develop strategies for STEM professionals to contribute to K-12 education. It is understandable to want scientific researchers to disseminate new knowledge into the science education system and become more involved in efforts to improve STEM education. Nevertheless, some observers have expressed concerns about the effectiveness of this approach. This is an appropriate time for NSF to re-evaluate this approach of mixing research projects with STEM education projects.

Achieving the right balance. The NSF's support for STEM education has historically been tugged in two different directions:

1. ***Supporting basic research and development.*** Basic research efforts includes studies of how students learn specific concepts and how teachers can best teach them, while basic development efforts include creation innovative instructional materials. Such basic R&D projects are well suited to individual university-based researchers, and they provide an essential foundation for improving STEM education. However, the research efforts are often only slowly translated into teaching practice and the development efforts rarely produce the kinds of integrated, whole-course instructional materials that teachers and students need in the classrooms.
2. ***Supporting development and implementation of scalable solutions.*** Some NSF programs have focused on creation and dissemination of solutions across the K-12 education system. Such efforts can have an enormous impact, but require larger and sustained efforts to achieve. Development efforts require initial creation of comprehensive materials, iterative cycles of testing and refinement, and active efforts to make the materials readily available to state and local authorities, either directly or through commercial publishers. At times, NSF has also faced political pushback due to the perception that it was endorsing particular projects over others.

We believe that the NSF's current funding portfolio is not optimally balanced to support the Nation's need. The emphasis on supporting development and implementation of scalable solutions should be substantially increased. This should ideally involve appropriation of additional resources for this work, but, if necessary, could appropriately involve some shifting of resources.

For various reasons, the time is right for an increased emphasis on development and implementation of scalable solutions. First, the state-led shared standards movements in mathematics, English language arts, and science have increased the prospects for excellent learning resources and materials to capture a larger share of the education market. Second, technology and digital tools make it possible to create rapid prototypes, gather data on their effectiveness, iteratively refine them and efficiently disseminate them. Third, the investments in the Interagency Education Research Initiative (IERI) program during the Clinton administration provided important lessons about scale-up.

Box 3-1. STEM Education Funding at NSF

The list below represents the largest categories of the NSF's STEM education funding for K-12 outreach, informal education, and teachers.[72] The amounts reflect funding levels for FY10 and under ARRA.

The Division of Research on Learning in Formal and Informal Settings

Runs investigator-driven grant programs focused on research and development related to resources, models, and tools for students and teachers; innovative technology experiences for students and teachers; research and evaluation on education in science and engineering; and research experiences for teachers.

- Discovery Research K-12 Program ($119 million): To advance pre-K through grade 12 student and teacher learning of the STEM disciplines through development and implementation of resources, models, and technologies.*
- Informal Science Education Program ($66 million): For informal science centers and new programming for public radio and television.*
- Research and Evaluation on Education in Science and Engineering ($46 million): To build the foundational knowledge base across the cognitive, learning, and STEM education sciences.

The Division of Undergraduate Education

Supports transforming undergraduate STEM education to meet the needs of the 21st century.

- Robert Noyce Fellowships ($55 million, plus $60 million in ARRA funding in FY09): For prospective mathematics and science teachers.
- Mathematics and Science Partnerships ($58 million, plus $2.5 million ARRA funding in FY09): For local programs that bring together university science and mathematics department in K-12 reform work with schools of education and local school districts.
- National Science Distributed Learning Program ($16 million), also known as National Science Digital Library: To establish a national network of learning environments and multimedia resources for STEM education at all levels, with open access for teachers.

The Division of Graduate Education

Supports U.S. graduate students and innovative graduate programs that prepare tomorrow's leaders in STEM.

- Graduate STEM Fellows in K-12 Education ($54 million), plus $2.5 million ARRA funding in FY09): Support for 875 graduate STEM Fellows who interact with teachers and students in K-12 schools and improve their communication and teaching skills, while enriching STEM education.

H-1 B Nonimmigrant Petitioner Fees, ITEST ($25 million)

Invests in K-12 activities that target the shortage of STEM professionals, with a current focus on information-technology-intensive careers.

Fourth, the commitment to STEM education in the administration, the States, the business communities, and the Nation as a whole presents the opportunity to engage in new, more concrete, and strategic partnerships with the Federal Government and with the corporate and philanthropic sectors.

Other Science Mission Agencies

Beyond the NSF, many of the other science mission agencies administer significant programs in K-12 STEM education.

NASA. NASA's K-12 STEM education and outreach portfolio includes several dozen programs focused on NASA's centers and grantee sites around the country. The programs are funded at a level of about $88 million annually,[73] the second largest investment by a science mission agency in K-12 STEM education. NASA programs include research experiences for teachers, design competition challenges for student teams, summer space camps and internships for students, and volunteer outreach opportunities for NASA employees in local schools.[74] The programs reflect a close alignment with NASA's agency-wide education strategy to inspire, engage, and educate through its programs.

While space is clearly an inspiring topic for students, the NRC studied NASA's K-12 programs and concluded that it is hard to judge their quality and impact.[75] As with other agencies' education programs, it is difficult to assess the NASA programs' alignment with broader educational goals. Evaluation of K-12 initiatives within the agency primarily involves documenting the geographic distribution of programs and the numbers of students, parents, and teachers who participate in the programs. The NRC panel charged with evaluating its K-12 education program in 2007 noted that the agency does not rigorously evaluate its education programs, nor does NASA consistently coordinate with other Federal agencies on its K-12 projects. This, the panel asserted, makes it difficult to know how effective the programs are.

NIH. NIH's K-12 STEM education funding of about $46 million per year focuses primarily on bringing the science content behind NIH-supported medical discoveries to the public, primarily through grants to informal education centers and science museums. Programs fund science exhibits for the general public and forums for public education and discussion about medical research and health topics. NIH also funds a curriculum-supplement series that teaches state-of-the-art content using research-based teaching strategies.[76] Examples include an oral health guide for elementary school students, a supplement on understanding alcohol (investigations into biology and behavior) at the middle school level; and an educational booklet on human genetic variation for high school students. For four years (2006–09), NIH also supported the Science Education Partnership Awards (SEPA) program to improve life science literacy through educational partnerships among biomedical and clinical researchers and K-12 teachers and schools, museums and science centers, media experts, and other educational organizations. The partners provided educational resources such as classroom curricula, instructional materials, mobile laboratories, workshops, films, software, and websites that give K-12 students, teachers, and the public a better understanding of the life sciences. Science centers and museums across the country used SEPA funding to develop stationary and traveling exhibits on fundamental biology and related topics.

Department of Commerce. The National Institute of Standards and Technology (NIST) and the National Oceanographic and Atmospheric Administration (NOAA) have K-12 education programs with budgets totaling $40 million in FY10. NIST offers a summer institute for middle school science teachers in local communities. NOAA offers a wide range of outreach, informal, and K-12 programs focusing on weather, climate science, and marine and coastal environments.[77] It was not until 2007 that NOAA received an agency-wide mandate for education through the America COMPETES Act. The Act required that NOAA develop a 20-year education plan and establish a new Education Council to lead the development of its education program. In a recent review of NOAA's education programs, the National Research Council noted that NOAA's summative evaluations have been carried out on a very small proportion of education activities, and there has been little consideration of evaluation that would enable the agency to recalibrate the entire portfolio to meet its goals effectively.[78] The NRC called for NOAA to develop an education plan, to define outcomes as measurable concepts that allow an assessment of whether a goal is being reached, and to clearly distinguish its impact on audiences from its activity outputs.

Department of Energy. The Department of Energy funds a range of programs focused on students and on teachers, with a budget of about $11 million in FY10. Some focus on students, such as the National Science Bowl; others, such as Academies Creating Teacher Scientists, focus on three-year programs of professional development for teachers in conjunction with national laboratories. In FY11, a new program that would devote up to $15 million to K-12 STEM education has been proposed called Re-ENERGYSE; it will develop curricula and laboratory equipment, on-line materials, and innovative approaches to provide energy literacy to the K-12 community.

Other agencies. The Department of Defense, the Environmental Protection Agency, the Department of Agriculture, the Department of Transportation, and the National Security Agency also fund various programs aimed at improving K-12 STEM education. The Department of Defense's largest program is called STARBASE, a $20 million program for students and teachers that emphasizes experiential applications, student interaction, and problem-solving experiments. Students and teachers visit military bases where they learn and apply knowledge in team inquiry and then add reasoning processes to build understanding of applied science, mathematics and technology. The Environmental Protection Agency supports environmental education for the public, and the Department of Agriculture funds school programs to introduce students and teachers to agriculture and agricultural careers. The Department of Transportation funds a general STEM education program focused on curriculum development and other activities related to transportation, and the National Security Agency funds the Mathematics Education Partnership Program.

Overall Federal K-12 STEM Education Portfolio

The current Federal investments in STEM education reflect an accumulation of programs and policies over several decades. The Obama administration views STEM education as a key priority; it has been moving to develop new blueprints for these investments to maximize their effectiveness and to improve coordination.

In support of these goals, we offer the following conclusions about prior patterns of Federal leadership and funding for STEM education.

1. **Federal investments in K-12 STEM education in recent decades have not been sufficiently strategic or coordinated.** The diversity of Federal projects and approaches to K-12 STEM education appears to have largely emerged without a coherent vision and without careful oversight of goals and outcomes. Efforts across agencies have not reflected a common set of goals and priorities for STEM education, and often this lack of coordination is apparent within agencies as well. Relatively little of the Federal funding has historically been targeted at catalytic efforts with the potential to transform STEM education. Where programs and practices have been demonstrated to be effective, there has been too little attention to replication and scale-up to disseminate their benefits more widely. And the Federal Government has not historically made good use of the potential contributions of the STEM communities in higher education, in industry, in Federal labs and research centers, and in nonprofit institutions.
2. **The Federal Government has historically lacked – and still lacks – sufficient leadership capacity dedicated to STEM education to accomplish the Nation's goals, as articulated by the President.** Neither the Department of Education nor the NSF has strong internal mechanisms that are sufficient to provide leadership and coordination. Secretary of Education Duncan has created the position of Special Assistant for Science, Technology, Engineering and Mathematics Education; we applaud this first step but note that the Department needs to put in place much greater capacity, such as an Office of STEM Education with adequate staff and budget. Similarly, the NSF has provided grants for investigator-driven projects in STEM education, and has provided some funding for scale-up of successful projects, but has not dedicated sufficient focus and resources to development and scale up of successful materials and programs in schools to meet the Nation's needs. In addition, historically it has not adequately coordinated with other agencies to drive policy changes based on its research findings.
3. **The amount of Federal funding historically dedicated to STEM education has been relatively small.** Across both the Department of Education and the science mission agencies, total Federal spending targeted to STEM education at the K-12 level was approximately $1.2 billion in FY10. This figure is small relative to total spending on K-12 education in the United States and to Federal spending on K-12 education. Moreover, the amount invested in R&D in education is miniscule relative to total spending on education; we estimate that it is roughly 0.2 percent. We believe that this amount is not adequate to drive the kinds of improvement needed in STEM education, nor is it consistent with the high priority the administration places on STEM education.

4. **Federally-funded programs and initiatives in STEM education have not historically been researched and evaluated in a manner that contributes to effective program development and policymaking.** The development of programs and policies should be coupled to systematic research and evaluation, ideally based on the most important outcome: student achievement in both the immediate term and the long term. In making investments, it is important to know what types of programs work well and which do not and for this knowledge to guide strategies and policies. In general, there has not been a systematic process for evaluating programs across agencies based on effectiveness and impact. (Earlier efforts to instill such a process under the National Science and Technology Council have had only modest effects to date.) Moreover, program evaluations within agencies have not contributed to an overall Federal strategy, nor have they been systematically designed and used to ensure the success of STEM initiatives in K-12 schools.

The Department of Education has now placed a higher priority on evaluation, and the Office of Management and Budget (OMB) has recently initiated discussions with agencies emphasizing the importance of evaluation.[79] OMB officials have noted the inconsistent manner in which evaluations are conducted and program outcomes are measured, as well as the failure of agencies to use evaluations to guide policy and program design.[80] OMB has initiated a process of making evaluations publicly available and accessible across agencies, and it plans to form an interagency working group to promote stronger and more consistent evaluation of Federal programs to learn what is effective and worthwhile. We support such efforts. We should note, however, that we do not mean to suggest that evaluation should necessarily consist of randomized controlled trials for most STEM education programs; instead, sound and appropriate methodological approaches for evaluating STEM education programs should be developed and used by all agencies.

While the lack of adequate knowledge on education program outcomes is not specific to the Federal Government, the Federal Government has a unique responsibility to address research challenges. Importantly, rigorous research on implementation of education initiatives is crucial to each of the recommendations we make in this report. For example, we need to know: the types of school settings in which a given initiative is most likely to succeed; the types of teacher professional development work that is most likely to improve student outcomes; how new, shared standards can be improved based on data gathered on student learning; what types of programs work best to recruit and prepare new STEM teachers; and which characteristics of schools and out-of-school programs enhance student learning and motivation.

It is also important that as a Nation we adopt a more realistic attitude toward all education policies and interventions. We cannot expect immediately to produce perfect programs through top-down decrees. Instead, we should learn by doing, study what we do, and adapt to the outcomes in classrooms, schools, and other settings to constantly improve STEM education initiatives and policy.

Leadership and Coordination within the Federal Government

The Obama administration has clearly signaled its intent to place much greater emphasis on K-12 STEM education programs throughout the Federal Government. What is needed is a more strategic, coherent, and coordinated set of programs across the agencies, as well as greater overall investment of dedicated funds. To achieve this coherence, new structures need to be created to ensure strong leadership, partnership, coordination, evaluation, and integration.

Improving STEM education will require a strong new partnership between the Department of Education and the National Science Foundation, supported by strategic leadership within each agency.

- The Department of Education is moving to play a more active role in K-12 STEM education under the administration's proposals. The Department has historically lacked significant internal leadership on STEM education. Currently, the Department has only one individual dedicated to creating and coordinating STEM education efforts within the Department and across agencies. It will need a new mechanism to provide leadership, strategic planning, and coordination. Ideally, this mechanism would be an Office of STEM Education with adequate personnel and budget to achieve critical goals.
- The National Science Foundation has historically had a major influence on K-12 STEM education, but in recent years it has not placed enough emphasis on projects that can make a widespread impact on schools and on education policy. The NSF should expand its strategic focus in STEM education and broaden its approach to maximize its impact. In addition to funding basic research and development on STEM teaching and learning, it should seek to drive systemic improvements in K-12 STEM education by funding potentially transformative technologies and materials and by focusing on how its projects can be scaled up through partnerships.
- The two agencies are complementary in their expertise. The Department of Education has strong ties to the Nation's schools and education policy but has lacked the strong ties to the STEM community that would allow it to incorporate scientific expertise into its projects. The National Science Foundation possesses the required staff expertise and the ties to the scientific community while lacking the systemic focus and levers for large-scale change that will be required to catalyze major changes in STEM education. A high-level partnership between the agencies could bridge this gap, driving systemic reform though innovative research and data-driven program evaluation. This partnership must be sustained by solid leadership, intertwined processes, and the alignment of major projects and resources.

In addition, the efforts of the many other Federal agencies engaged in STEM education activities are not well coordinated. The Federal Government lacks an effective mechanism to align the strategy, execution, and evaluation of the STEM education activities across agencies. There have been *ad hoc* efforts, such as a temporary education subcommittee of the National Science and Technology Council (NSTC).[81] A standing NSTC Committee on STEM Education could help ensure a more coherent approach.

We believe these activities will require a significant number of new full-time employees in professional staff positions and a corresponding number of administrative support positions, whereas currently only one professional full-time employee focuses on STEM at the department.

RECOMMENDATION 3-1. LEADERSHIP WITHIN THE DEPARTMENT OF EDUCATION

1. **The Department of Education should continue its increased focus on STEM education through the creation of transformative programs and through targeting of ESEA funds to STEM education. The Department also should take complementary steps, such as increasing preference points for STEM- related initiatives in its Race to the Top and Investing in Innovation funds.**
2. **The Secretary of Education should create a strong mechanism to ensure leadership and coordination for STEM programs within the department – for example, an Office of STEM Education, with adequate staff and budget. Responsibilities should include developing overall strategy for STEM education in the department; coordinating grants and programs focused primarily on STEM education; ensuring that other programs in the department with STEM education components reflect best practices for improving STEM education; working with other Federal agencies to ensure that their efforts align with Federal strategy and national needs; and helping make scientific resources of other Federal agencies accessible and useful to schools, school districts, and states.**

RECOMMENDATION 3-2. LEADERSHIP WITHIN THE NATIONAL SCIENCE FOUNDATION

1. **The National Science Foundation should expand its strategic focus in STEM education. In addition to funding basic research and development, it should seek to drive systemic improvement by funding potentially transformative efforts aimed at developing educational technologies and fully integrated learning resources in STEM subjects and by partnering with other agencies to scale up the most effective projects for improving STEM education in schools.**
2. **The NSF Director should create a strong mechanism to ensure leadership and coordination for STEM education programs within the agency. Responsibilities should include developing an agency-wide strategy for STEM education; ensuring that NSF projects related to K-12 STEM education identify and expand best practices for improving STEM education; and coordinating with other Federal agencies.**

To increase its capacity for and engagement with the STEM community, the Department of Education could consider such practices as using Interagency Personal Agreements (IPAs) to bring STEM experts into the department for several-year periods (a practice routinely used by the NSF and the Defense Advanced Research Projects Agency) and holding regular conferences on STEM education, to engage the STEM education and STEM professional communities.

OSTP should request that NSTC develop, within six months, an integrated cross-agency strategic plan for Federal investments aligned with the President's goals for STEM education. In addition, the President and the Office of Management and Budget should request that all agencies include in their reauthorizations specific language that enables them to more easily and more quickly respond to national STEM education goals and to collaborate with other agencies to meet those goals. We note that the recommendation above is similar to provisions in the reauthorization of the America COMPETES Act recently passed by the House of Representatives.[82]

Recommendation 3-3. A Partnership between the Department of Education and the National Science Foundation

The Department of Education and the National Science Foundation should enter into a high-level partnership to improve STEM education, using their complementary expertise to engage the education community and STEM communities. The partnership should be sustained through the designation of committed leaders, and through the intertwined processes, projects and resources.

The agencies should also promote the practice of research and development, data collection, analysis, and evaluation throughout STEM education entities so that policies are ultimately based on solid empirical foundations. At the same time, recognizing that it may take a decade or more to definitively prove the effectiveness of programs, the agencies should not be afraid to act boldly based on best available professional judgment.

Recommendation 3-4. Coordination across Federal Agencies

The Federal Government should create, within the National Science and Technology Council, a standing Committee on STEM Education co-chaired by the Department of Education, the National Science Foundation, and the Office of Science and Technology Policy.

The Committee should create a Federal STEM education strategy, should update it annually, and should work with OMB to ensure that STEM education programs within the Federal Government are aligned.

The Committee also should help develop appropriate and consistent data planning, collection, and analysis techniques for evaluating programs across agencies. It should foster the development of a high- quality research agenda designed to ensure that Federal programs learn and improve over time.

Advice and Support from Outside Government

As the Federal Government takes on a more vigorous role in STEM education in K-12 schools, it also needs to engage and use the expertise that resides outside its boundaries in academia, nonprofits, state councils, and the private sector.

The Nation needs a structure and mechanism for stakeholders and experts outside of the Federal Government to provide independent, objective, and data-driven guidance and advice to advance STEM education goals. This requires an entity that would review progress toward the critically important goals of moving the Nation to "the top of the pack" internationally and closing the achievement gap in STEM education.

Most states have STEM leadership advisory councils that frequently report to their governors. These councils include representatives from K-12 education, higher education, government laboratories, informal science education institutions, and local industry. These advisory groups are another important resource for understanding states' challenges and resources and for focusing state resources on STEM education initiatives. Connecting with these councils and sharing data, information, and strategies will increase the overall leadership and coherence of STEM education efforts.

Through its "Educate to Innovate" initiative, the administration has reached out to the private sector to play a more active role to support education goals across the country. While these private sector financial and in-kind commitments are needed and welcome, they are only an initial step toward forging a deeper, stronger, and longer-term commitment to improving STEM education for all students. These private sector groups also should help make the case to the Nation's youth that STEM careers hold rewards, excitement, and professional satisfaction, and that the culture of STEM professions will be accepting and encouraging to them.

A Commission involving businesses and universities could provide important input and feedback to the Federal Government about STEM education. Moreover, we believe that such a Commission would be most effective if done jointly with the states.

RECOMMENDATION 3-5. A COMMISSION ON STEM EDUCATION

We recommend that the President create an independent Commission on STEM Education, established in conjunction with the National Governors Association and with a life of at least ten years, to monitor and promote progress toward improving STEM education.

The Commission should report once every two years on progress toward the goals of moving the Nation to "the top of the pack" internationally and closing the achievement gap in STEM education. It also should promote the importance of STEM education and propose strategies to coordinate and improve efforts by the Federal Government, states and localities, private industry, nonprofit science organizations, and universities.

We estimate that the Commission will require modest funding for professional staff and operations. Private-sector groups, such as the newly formed Change the Equation, could play a key role in helping to organize and marshal corporate and philanthropic support for the Commission's activities.

IV. Shared Standards and Assessments

Chapter Summary

The Nation will benefit from shared standards in STEM subjects in K-12 to prepare students to compete in a global economy. We also require high-quality assessments that truly reflect what students need to learn and are learning, as these will further progress in STEM education and allow educators to improve their approaches and materials. The recent state-led effort that has developed common core standards in mathematics – and the ongoing effort to develop new standards in science that could be adopted or aligned to by all states – signals an exciting new era for STEM education. Shared standards create the need and opportunity to provide better materials and professional development for teachers and to lay the groundwork for fair and valid assessments that measure what students have learned and benchmark U.S. performance against that of other countries. Continued work at the state level is vital, as is support from the STEM professional community to advance new science standards. The Federal Government should vigorously support the assessments and teacher professional development necessary to seize this historic moment and opportunity for shared standards.

Introduction

For the K-12 public education system in the United States to achieve excellence, it must take the approach that underlies excellence in any system: It must have clear, consistent goals and measure progress toward them. In education, a critical goal is that students achieve a certain level of understanding and skills. **Standards** define the level; **assessments** measure whether students are achieving it.

The fact that education is largely the responsibility of states and localities in the United States has certain advantages. It promotes local responsibility for education, allows responsiveness to local needs, and enables valuable experimentation with new models and approaches. However, this system also poses challenges, given that students will live and work in settings that are increasingly national and international.

Most important for STEM education, the education system in the United States has produced a situation in which 50 different sets of standards exist for mathematics and for science.[83] This fragmentation makes the system less coherent, less effective, and less efficient. Students, states, and the Nation would be better served by standards and assessments that are broadly shared, for several reasons. First, the Nation increasingly competes in a global economy in which science and technology are central. The principles and practices of mathematics and science do not differ across geographic boundaries: the laws of nature are the same in Biloxi, Boise, and Boston – and, importantly, in Beijing, Bangalore, and Brazil. To be broadly prepared to participate and compete in a world economy driven by STEM-based innovations, our students need a common, strong foundation of knowledge and skills. Second, families in the United States frequently move across state lines, and differing standards create difficulties for students and teachers alike. Third, shared standards and assessments will facilitate cross-state comparisons to help schools identify and implement the country's best educational practices. Shared standards will promote accountability – of students, teachers, schools, and systems – and will help to identify both strengths and weaknesses in the system. And fourth, shared standards will create a large and more unified marketplace for educational goods and services. Expanding and defining that marketplace will provide the incentives needed to drive creativity and innovation in the development of state standards and assessments, as well as in the development of high-quality instructional materials, educational technologies, and professional development for teachers and school leaders.

In addition to being shared, standards and assessments should reflect what students truly need to know. Standards should be aimed not at memorizing facts but at achieving a combination of factual knowledge, conceptual understanding, procedural skills, and habits of thought. Assessments should measure the extent to which students have mastered this full range of learning. Although such high-quality assessments are more expensive to develop and score, they will ultimately save money by ensuring that other investments throughout the education system achieve their aims.

Shared standards, however, do not imply Federal standards. In many sectors and industries, consistent standards that span the country or the world provide a common base of knowledge, understanding, and expectations that guide individual strategies and policies. Individual actors, whether states and localities or firms, make decisions guided by such standards and innovate upon them. Some states may adapt the shared standards to meet their needs. It is important, however, that these standards align with each other; almost every country that does well in international comparisons of STEM education has shared standards of some sort.

In this chapter, we provide a brief history of efforts to develop STEM education standards in the United States. This is followed by a discussion of assessments that are fair and valid and measure important knowledge and skills in ways that provide valuable feedback to students and teachers while also providing data that are useful for school accountability. We then make recommendations concerning the critical role the Federal Government can play in encouraging the work of the state-led shared standards movement.

We emphasize that the Federal Government does not itself define the curricula, standards, or assessment used in public schools in the United States. This is a responsibility of state and local authorities throughout the Nation. The movement toward common standards is an effort originating from and led by the states. The Federal Government's role should be to support these state-led efforts.

Initial Efforts at Standards

Mathematics and science are just two components of a complex K-12 education system. And while mathematics and science are linked together in common parlance and are intertwined in practice, they are often treated quite separately in our education system. The historical emphasis in public schools on "reading, writing, and arithmetic" has contributed in part to the dominance of mathematics and English language arts, especially in grades K-8. More recently, high-stakes annual testing required by the No Child Left Behind Act in these subjects, beginning in grade 3, has reinforced that emphasis. Science and technology, while important, have not played a prominent role in the curriculum, especially in the early grades.

The differences between the mathematics and science communities also have led to differences in the development of mathematics and science standards over the past 25 years.

Mathematics Education Standards. The National Council of Teachers of Mathematics (NCTM) developed the first K-12 mathematics education standards for the country in 1989. Around the same time, the National Research Council created the Mathematical Sciences Education Board (MSEB), which brought together various mathematics research and education constituencies to focus on understanding mathematics learning at the K-12 and undergraduate levels and improving mathematics education. MSEB used several mechanisms (convocations, symposia, focused sessions at annual professional meetings, and publication of reports) to engage the full mathematics community and policy makers in discussions about what mathematics education should be in the nation's schools and universities.

After the mathematics standards were released in 1989, the states put forth significant effort throughout the 1990s to adopt, adapt, or develop their own school mathematics standards; they also developed new curricula, instructional materials, assessments, and teacher professional development. This period also saw what came to be called the "math wars," pitting traditionalists who favored basic skills such as arithmetic against reformers who favored mathematical understanding. Those disputes have subsided in recent years with the more widespread recognition that effective education requires both aspects; it also requires much more.

In 1995, NCTM recognized that its Standards would need to be periodically examined, evaluated, and revised to remain relevant. After an extensive drafting and review process, in 2000, NCTM published "Principles and Standards for School Mathematics," or PSSM. The Principles established a foundation for school mathematics programs by considering the broad issues of equity, curriculum, teaching, learning, assessment, and technology. The Standards for school mathematics describe an ambitious and comprehensive set of goals for mathematics instruction, and the basic skills and understandings that students will need to function effectively in the 21st century.

Similar issues were expressed in the report of National Mathematics Advisory Panel in 2008,[84] which concluded that:

- The mathematics curriculum in Grades Pre K-8 should be streamlined and should emphasize a well-defined set of the most critical topics in the early grades.
- Standards and curricula should make use of what is known from rigorous research about how children learn, especially by recognizing (a) the advantages for children in having a strong start; (b) the mutually reinforcing benefits of conceptual understanding, procedural fluency, and automatic (i.e., quick and effortless) recall of facts; and (c) that effort, not just inherent talent, counts in mathematical achievement.
- Our citizens and their education leadership should recognize mathematically knowledgeable classroom teachers as having a central role in mathematics education and should encourage rigorously evaluated initiatives for attracting and appropriately preparing prospective teachers, and for evaluating and retaining effective teachers.

- Instructional practice should be informed by high-quality research, when available, and by the best professional judgment and experience of accomplished classroom teachers. High-quality research does not support the contention that instruction should be either entirely "student centered" or "teacher directed." Research indicates that some forms of particular instructional practices can have a positive impact under specified conditions.
- NAEP and state assessments should be improved in quality and should carry increased emphasis on the most critical knowledge and skills leading to Algebra.[85]

Science Education Standards. Efforts to describe a body of knowledge to guide science education emerged in the mid-1980s. The American Association for the Advancement of Science (AAAS) led the initial effort and published "Science for All Americans" in 1989 and "Benchmarks for Science Literacy" in 1993. These products of AAAS's [86] were envisioned as -term initiative to advanceProject 2061 literacy in science, mathematics, and technology. The benchmarks were not intended to be adopted as standards, but they were and continue to be influential in shifting public and professional thinking about the purposes of science and mathematics education.

In the early 1990s, a parallel and complementary effort began, managed by the National Research Council (NRC), to produce science education standards. This process was furthered by the Goals 2000: Educate America Act that was signed into law in 1994, which established standards in mathematics, science, and other subjects, in order to meet the national goal of U.S. students being first in the world in science and mathematics achievement by 2000 – a goal that was manifestly not achieved. In early 1996 the NRC published the National Science Education Standards, which included content standards for science as well as standards for teaching, professional development, assessment, science education programs, and education systems.

In the years that followed, states adopted science standards using elements of both the AAAS's and the NRC's work and with varying levels of fidelity to the source documents. These standards efforts, however, did not always have the effects their authors intended. The standards were supposed to provide templates that the states could use to produce their own sets of high, unified standards that would set ambitious learning goals for students. In turn, the standards would motivate the creation of high-level instructional materials and assessments that would improve student learning in science, mathematics, and technology.

Instead, the states produced sets of standards that were often quite different from the national standards and from each other's, with greatly varying content, structure, and quality. Many of the state-level standards emphasized low-level skills and large bodies of factual content rather than the high-level abilities and central concepts emphasized in the national standards.

In most states, there are so many standards it is difficult to imagine that teachers can cover the full range of information – much less impart understanding – in a given school year. In one large state, there are 15 sets of knowledge and skills for mathematics (with subsections under each set) and 10 sets for science – in kindergarten! The standards increase in number and complexity with each grade level, so that in grade 3 students are expected to, for example, "connect Grade 3 science concepts with the history of science and contributions of scientists."

The scientific community itself often contributes to the problem, with each discipline angling to include specific details in the standards from its domain. In one large state, high school biology has 67 specific standards, including "Students know the role of the endoplasmic reticulum and Golgi apparatus in the secretion of proteins." The distinguished cell biologist Bruce Alberts, former President of the National Academy of Sciences and a member of the PCAST STEM Education Working Group, sees little purpose in that standard. John Holdren, Assistant to the President for Science and Technology, Director of OSTP, and co-chair of PCAST, observed at a recent public meeting of PCAST that he has worked as a scientist for 40 years without knowing what the endoplasmic reticulum is. This example is merely illustrative, but it demonstrates the problem with state standards consisting of long lists of factual material that students are supposed to memorize.

Educational publishers, in an effort to produce materials that could be used in as many states as possible, have produced textbooks crammed so full of facts that the central concepts of a field are difficult to discern and apply. And annual assessments, despite some promising experimentation, have focused largely on low-level skills and factual recall rather than the capabilities students need in the 21st century. This could account for our students' lackluster performance on national and international assessments of mathematics and science. On the TIMSS assessments, for example, which emphasize basic skills and factual knowledge learned in classrooms, U.S. students rank in the middle of the pack. On the PISA tests, which focus on 15-year-old students' application of conceptual understanding and scientific thinking to problems and decisions, American students perform in the bottom ranks.

Rethinking Standards

Over the last decade, research has led to a deeper understanding of how people learn and has provided powerful insights into the learning of mathematics and science from the novice to expert levels. Committees of experts convened by the National Research Council have reviewed knowledge about mathematics and science learning and about the abilities most important to cultivate in students. Such research is now informing new efforts to rethink standards in mathematics and science.

In addition, important lessons emerged from initial efforts to create shared standards that have shaped a new approach. Past standards lacked a sense of prioritization of what was most important for students to learn and for educators to teach, which made them difficult to put into practice. New efforts on the part of NRC and other organizations have taken into account this history and have responded to it.

In 2001 an NRC panel identified the key strands for effective mathematics education. Those strands balance the need for skills to carry out mathematical procedures with the need for conceptual understanding and complex reasoning (see Box 4-1). The idea is to expose students to elegant concepts and patterns in mathematics so they can understand its beauty while also teaching them skills they need to apply those concepts, enabling them to see the utility and relevance of mathematics. In 2007 an NRC panel published a parallel set of strands for science education (see Box 4-2). These strands transcend the emphasis on low-level factual recall found in many science classes today to include the skills needed to solve complex problems, work in teams, and interpret and communicate scientific information.

Box 4-1. Strands for Mathematics Education

In 2001 a National Research Council committee sought to transcend the historical debate about which was more important in mathematics education – understanding or skills – by identifying five key strands in effective mathematics education:

1. *Conceptual understanding* – comprehension of mathematical concepts, operations, and relations
2. *Procedural fluency* – skill in carrying out procedures flexibly, accurately, and appropriately
3. *Strategic competence* – ability to formulate, represent, and solve mathematical problems
4. *Adaptive reasoning* – capacity for logical thought, reflection, explanation, and justification
5. *Productive disposition* – habitual inclination to see mathematics as sensible, useful, and worthwhile, coupled with a belief in diligence and one's own efficacy.

Source: National Research Council (2001), *Adding it Up: Helping Children Learn Mathematics,* Washington, DC: National Academies Press.

Box 4-2. Strands for Science Education

In 2007 another National Research Council committee examined how science is learned and the critical stages in children's development of scientific proficiency in science and defined four key strands of science education, recommending that students should:

1. Know, use, and interpret scientific explanations of the natural world. Students acquire facts and conceptual structures that incorporate those structures and use them to understand many phenomena in the natural world.
2. Generate and evaluate scientific evidence and explanations. Students have the knowledge and skills to build and refine models based on evidence, including designing and analyzing empirical investigations and using empirical evidence to construct and defend arguments.
3. Understand the nature and development of scientific knowledge. Students recognize that science is a particular kind of knowledge with its own sources, justifications, and uncertainties, and that predictions or explanations can be revised on the basis of new evidence or a new conceptual model.
4. Participate productively in scientific practices and discourses. Students understand the norms of the practice of science and how to participate in scientific debates or adopt a critical stance, and they are willing to ask questions.

Source: National Research Council (2007), *Taking Science to School*, Washington, DC: National Academies Press.

In the mid-2000s several research and policy efforts directed at education reached the same conclusion: student achievement at all levels would benefit from fewer, clearer, and higher standards. These efforts have been coordinated and integrated with the NRC's work to shape new standards.

In mathematics the National Council of Teachers of Mathematics, which developed the first national mathematics standards effort, began a "focal points" project to describe important mathematical topics for each grade level from pre-kindergarten to eighth grade.[87] The idea was that these areas of instructional emphasis could then serve as organizing structures for curriculum design and instruction at and across grade levels. Also, Achieve — an independent, nonpartisan, nonprofit organization created in 1996 by the nation's governors and corporate leaders to help states raise academic standards and graduation requirements, improve assessments, and strengthen accountability – became active in trying to move states toward more coherence in mathematics and language arts through its America Diploma Project and K-12 Benchmarks.

Recently, the Carnegie Corporation of New York and the Institute for Advanced Study looked in depth at mathematics and science education – including content, teaching and learning, and schools and school systems. These organizations concluded that a huge opportunity exists to transform American education by enabling higher levels of mathematics and science learning for all American students. Their report called for "common standards in mathematics and science that are fewer, clearer, and higher, coupled with aligned assessments."[88]

Shared Standards Movement

These calls for reform have added impetus to an important new effort led by the states to establish shared standards in specific subjects. In 2009 the National Governors Association (NGA) and the Council of Chief State School Officers (CCSSO), working with Achieve, Inc., and with the college entrance testing organizations ACT Inc. and the College Board, initiated the Common Core State Standards Initiative, focusing on mathematics and English-language arts.[89] Teachers, school administrators, and subject matter experts developed the standards, with opportunities for input from a wide range of stakeholders. Final K12 standards in mathematics and English-language arts were released in June 2010. As of the publication date of this report, 36 states and the District of Columbia have agreed to adopt – or adapt their standards to align with – the common core standards in mathematics and English language arts. Adoption and alignment of standards can in turn be expected to drive changes in instructional materials, assessments, and professional development.

A complementary effort is under way in science. A committee of the National Research Council is producing a framework that will be used to develop the new science standards. [90] The framework is scheduled to be completed by early 2011, with the development of full K-12 science standards by the organization Achieve, Inc., through an iterative process involving states and stakeholders and in consultation with professional groups, to be completed by the end of 2011. It will be very important for state-level groups such as the National Governors Association and Council of Chief State School Officers to promote the standards as soon as they are completed, so that states can adopt them and begin the work of helping students reach these levels of achievement.

Technology and Engineering

To date, most states have no standards in technology and engineering subject areas.[91,92] Yet, a basic understanding of technology and engineering is important if our children are to contribute to and compete in a rapidly changing society and an increasingly interconnected global community. We support recent efforts to develop assessments of technological learning[93] and to incorporate aspects and examples of technology and engineering (and design principles) both in mathematics standards and in the framework for science standards.[94 95]

We also believe that there is an urgent need for well-designed courses in technology and engineering, with high-quality instructional materials, particularly in high schools. Computer-related courses should aim not just for technological literacy, which includes such utilitarian skills as keyboarding and the use of commercial software packages and the Internet, but for a deeper understanding of the essential concepts, methods and wide-ranging applications of computer science. Students should gain hands-on exposure to the process of algorithmic thinking and its realization in the form of a computer program, to the use of computational techniques for real-world problem solving, and to such pervasive computational themes as modeling and abstraction, modularity and reusability, computational efficiency, testing and debugging, and the management of complexity. Where feasible, active learning, higher-level thinking, and creative design should be encouraged by situating new concepts and techniques within the context of applications of particular interest to a given student or project team.

Over the past few years, a number of organizations have begun to formulate guidelines, principles, and proposed standards for the incorporation of technology-related subject matter within the K-12 curriculum. The Association for Computing Machinery (ACM) has developed a model K-12 computer science curriculum[96] that presents a framework within which school districts and state departments of education can revise their curricula and standards to include key elements of computer science and computational thinking. In 2005, the Computer Science Teachers Association (which evolved from the ACM's K-12 Task Force) produced a report that proposed a set of principles and guidelines aimed at improving high school computer science education.[97] In a project sponsored by NSF's Directorate for Computer and Information Science and Engineering (CISE), the College Board is developing a new Advanced Placement course designed to introduce students to the central ideas of computing, computer science, and computational thinking. [98] Although each of these efforts has had a somewhat different focus, they are generally reflective of a shared perspective on many of the most important aspects of K-12 education in IT-related areas.

Engineering education must motivate students to understand concepts and principles, yet it must also engage them to actively design and build and to solve challenges they encounter. The National Academy of Engineering recently evaluated the Nation's needs with respect to engineering education in K-12 and has helped highlight the key concepts and abilities that students should acquire.[99] The NAE report also presents general principles for K-12 engineering education, including the emphasis of engineering design; the incorporation of developmentally appropriate mathematics, science, and technology knowledge and skills; and the promotion of "engineering habits of mind." Programs such as Project Lead the Way have already made strides in integrating engineering and design principles into K-12 curriculum and form a useful basis for future efforts.

The essential components of technological and engineering education will evolve over time, and it is important that the Nation continue to consider the skills and knowledge that K-12 education should provide to all students. Scattered examples of "school to work" programs around the United States have demonstrated the value of using technical education and workplace linkages as a way to improve the academic preparation of all students, whether they are headed directly into the workforce or into two- year or four-year colleges.

Assessments

Standards are only a part of the equation. To implement standards well, schools require leadership, high- quality instructional materials aligned to standards, and well-prepared teachers. We address these needs in our recommendations below. Beyond these elements, the Nation also requires a major complementary investment in high-quality, meaningful assessments. Standards are only valuable if students, teachers, and schools have effective ways to tell if they are achieving the standards – and to be held accountable by parents, districts and States. In addition, the true meaning of a standard is often unclear until the corresponding assessment has been defined.[100] (See Box 4-3.)

There are several types of assessments: **Formative tests** are used in the process of teaching a concept or skill. Teachers can use formative assessments to shape what content to provide and how to teach it; students can use formative tests to assess their own progress. **Summative assessments** measure student learning retrospectively, after a unit has been taught or a concept completed, so that teachers can see whether and which students grasped and retained the material they covered. Both formative and summative assessments are given within the context of courses and inform student work and teacher practice. Teachers typically drive how and when they give these assessments (with exceptions such as those given for AP and IB courses and by the Algebra II consortium). **End-of-year summative tests** are generally given to see whether individual students have achieved particular levels of learning and how well groups of students are achieving. By and large, they do not inform students about their performance in ways that can affect the learning process during the course of the school year. They are used to benchmark and track overall progress. Importantly, they can have a powerful effect on what school systems demand from teachers, particularly when the results lead to rewards or punitive consequences. Under the No Child Left Behind Act, for example, schools are assessed on a proficiency scale based on student performance and are required to make annual progress for all student groups, with all students expected to be at proficiency or above by 2014. Finally, **research assessments** are used to inform the development of curricula, instructional materials, and teaching methods.

The administration recognizes the need for new high-quality assessments and has allocated $360 million in FY10 for a competition in which consortia of states develop new assessments for college- and career- ready standards for mathematics and English language arts. Two consortia of states – one involving 31 states, the other 26 states – have been selected to develop these assessments. The emergence of groups of states self-organized to support the development of shared assessments is a departure from the traditional state-by-state assessment development process in which each state contracts with companies to develop state-specific accountability measures for the grades tested (generally each year in grades 3 through 8, and once in high school). The administration also has allocated $30 million for high school endof-course assessments. Plans include at least three board examination systems in lower-division high school grades and five in the upper division – in English, mathematics, science, history, and three career and technical occupational groupings.

The Need for Clear, High-Quality Assessments. A useful assessment makes clear whether a student has met a standard and reveals what teachers should be covering to help their students advance.

BOX 4-3. ONE STANDARD, MANY WAYS TO ASSESS

The true meaning of a standard often only becomes clear when one specifies the corresponding assessment. *A Practical Guide to Improve Instruction* by Paul Bambrick-Santoyo illustrates this point nicely by giving alternative question that might be used to assess a single state standard.

The standard reads: "Understand and use ratios, proportions and percents in a variety of situations." (New Jersey's Core Curriculum Content Standards for Mathematics, Grade 7, 4.1.A.3.) Six different assessment questions that could be said to be "aligned" to the standard are:

1. Identify 50% of 20.
2. Identify 67% of 81.
3. Shawn got 7 correct answers out of 10 possible answers on his science test. What percent of questions did he get correct?
4. J.J. Redick was on pace to set an NCAA record in career free throw percentage. Leading in the NCAA tournament in 2004, he made 97 of 104 free throw attempts. What percentage of free throws did he make?
5. J.J. Redick was on pace to set an NCAA record in career free throw percentage. Leading into the NCAA tournament in 2004, he made 97 of 104 free throw attempts. In the first tournament game, Redick missed his first five free throws. How far did his percentage drop from before the tournament game to right after missing those free throws?
6. J.J. Redick and Chris Paul were competing for the best free-throw shooting percentage. Redick made 94% of his first 103 shots, while Paul made 47 of 51 shots.

a. Which one had a better shooting percentage?
b. In the next game, Redick made only 2 of 10 shots while Paul made 7 of 10 shots. What are their new overall shooting percentages?
c. Who is the better shooter?
d. Jason argued that if Paul and J.J. each made the next 10 shots, their shooting percentages would go up the same amount. Is this true? Why or why not?

Source: Paul Bambrick-Santoyo. (2010). *Driven by Data: A Practical Guide to Improve Instruction*. San Francisco: JosseyBass, pp. 6-7.

But the appropriate formative and summative assessments for a given standard are often not obvious; nor are such assessments necessarily aligned with what is measured in current accountability exams.

For example, in New Jersey, one standard stipulates that students in grade 7 should be able to "*understand and use ratios, proportions, and percents in a variety of situations.*" In his book *Driven by Data*, Paul Bambrick-Santoyo shows how this single standard can correspond to a wide range of possible levels of understanding of ratios, proportions, and percents, depending on how the standard is assessed (see Box 4-3). In short, the meaning of a standard only becomes clear in light of the difficulty, scope, and design of the questions used to assess it. Bambrick-Santoyo writes that "if teachers were given this standard without clarification and commentary, no one could fault them" if they taught only the skills needed to answer an easy problem, even if the end-of-year state test demanded skills necessary to answer a very difficult problem.

Assessments require just as much attention and clarity as the standards themselves. To effectively teach STEM and raise student achievement to the level of standards and beyond, schools and teachers also need much more support in acquiring, delivering, and learning from assessments. The assessments must reflect what we want students to learn. As important, assessments need to be fair, valid, and transparent to maintain the public's trust. It is crucial, moreover, that teachers agree to, embrace, and fully understand assessments so that they fulfill their purpose. School and school district leaders should also have an understanding of the assessments and their usefulness. They should know how to use the data yielded by assessments for reporting, management, and administration purposes.

Assessments aligned with new state standards also need to foster high-quality teaching rather than discourage it. Such assessments should measure higher levels of thinking and reasoning as well as students' content knowledge and skills. They need to help all students achieve at high levels while not holding back students who want to explore a subject more quickly or at a deeper level. Thus, when teachers aim to increase student scores on these assessments, they should foster all the types of learning that the standards emphasize – not merely the factual recall aspects of learning that are by far the easiest and least expensive to test. A good assessment encourages quality teaching and learning. This is no small feat given that excellence in STEM education means cultivating in students not simply the ability to answer predictable questions, but the capacity to pose probing questions and to figure out methods of answering those questions.

Most current assessments fail to meet these goals. Observers and educators note that they tend to overrepresent low-level skills and factual recall, which may lead teachers to drill students on specific skills and facts they need for the test. Current tests often do a mediocre job of measuring the understanding and application of core concepts and principles, and they typically neglect the higher-order reasoning, problem-solving skills, and mathematical and scientific creativity that students need for college and for their careers.

The reason for the widespread use of these annual mathematics and science assessments is clear: Assessments that measure the ability to apply concepts and reasoning are harder to write and are more expensive to administer and score. Currently, average annual state-assessment costs for mathematics, reading, and writing are about $20 per student. For a moderately sized state, a four-year testing cycle costs about $52 million.[101] The national costs for annual state assessments required by the No Child Left Behind Act are estimated to be in excess of $800 million.[102] [103] Most of the costs are borne by the states, with the Federal

Government providing about $400 million annually to support required testing programs. Innovative, higher quality assessments are estimated to cost considerably more than traditional assessments – perhaps as much as $56 per student per year for mathematics, reading, and writing, with lower costs estimated if consortia of states organize to develop and administer assessments (see Box 4-4).

Inexpensive mathematics and science assessments, however, represent a false economy. They can undermine mathematics and science programs when teachers feel obligated to teach to poor tests. An annual educational investment of as much as $10,000 or more per student should not be driven by assessments costing $20 per student or less. Moreover, technology-based assessments have the potential to make such high-quality assessments available at a lower cost.

Currently, most states contract with testing companies to produce assessments that are aligned with their own specific standards. Developing the assessments accounts for about 25 percent of the total cost; the remainder of the cost is in administering the tests. This state-by-state development process is inefficient and costly and impedes comparison of results across states. Shared standards and assessments will make the process more efficient and could be used to refocus assessments on higher level skills and knowledge. Online assessments could drive costs even lower. (Box 4-4 summarizes cost models for new assessments.) Although the cost of developing higher quality assessments will initially be higher, the information gathered from such assessments will be more valuable for students and teachers, as well as for purposes of accountability.

The critical goal for new assessments is that they should truly reflect what we want students to learn. When teachers teach with the goal of improving test scores (that is, when they "teach to the test"), it should be the case that they are emphasizing the higher level skills that all students will need for success as adults in today's world. Creating assessment systems that meet this goal will require vigorous support for long-term, high-quality research in actual schools and classrooms focused on producing the feedback needed for successive improvements of assessment systems.

Benchmarking. In addition to state assessments tied directly to common standards, it is important to increase participation in key national and international assessments, including the NAEP, the TIMSS, and the PISA. Participation in these tests will provide crucial information that will allow states and the Nation to benchmark U.S. performance against high-performing nations. PISA, because of its emphasis on conceptual understanding, critical thinking, and problem solving, may be especially important as a method to measure progress in STEM education.

Role of Technology in Assessment. The Department of Education has recognized that the collection and use of data is a critical foundation for improvement, including it in the Department's four assurances. One reason that teachers, schools, and school systems collect and use too little data is that the process is so difficult. Creating, administering, and evaluating assessments is time-consuming. Here, we believe that technology can and must play an important role.

The recent National Education Technology Plan noted[107] that: "Most of the assessment done in schools today is after the fact and designed to indicate only whether students have learned.

BOX 4-4. COST APPROXIMATION MODELS TO DEVELOP & ADMINISTER NEW ASSESSMENTS

Today most states contract with testing companies to produce annual assessments that are aligned with their own specific content standards. Annual national costs are estimated to be more than $800 million. This distributed process is costly. It also impedes comparison of assessments results across states. Shared standards and assessments will make the process more efficient and could be used to refocus assessments on higher-level skills and knowledge for all students.

The costs for higher-quality assessments that truly reflect the goals of the new standards will initially be higher. The information gathered, however, would be more valuable for students and teachers and for purposes of accountability. It would also better reflect the content that teachers should teach and that students should learn in math and science.

The precise costs are difficult to know with certainty, but a recent analysis provides some estimates. If individual states developed their own assessments, the per-student cost would rise from about $20 to about $56 per student per year for mathematics, reading, and writing combined.[104] The cost models indicate that consortia of states (ranging from 10 to 30 states) working together to develop higher quality assessments could both increase the quality of assessments and drive the costs down by 30 percent (and online assessments could drive the costs even lower, as will scoring methods that involve teachers). [105] [106] While the assessments would still be more expensive than they are today, better assessments would drive the kind of teaching and learning the Nation needs.

Per-Pupil Cost of Higher Quality Assessments, Per Year, by Consortium Size
Assessment costs for mathematics, reading, and writing

Size	Year 0	Year 1	Year 2	Year 3	Average Per Pupil
1 State	$8.93	$52.07	$51.97	$54.05	$55.67
10 States	$1.33	$40.54	$41.83	$43.51	$42.41
20 States	$1.00	$38.86	$40.26	$41.87	$40.66
30 States	$0.86	$37.10	$38.49	$40.03	$38.83

Source: Topol, B., Olson, J., and Roeber, E. (2010). *The Cost of New Higher Quality Assessments: A Comprehensive Analysis of the Potential Costs for Future State Assessments.* Stanford, CA: Stanford University, Stanford Center for Opportunity Policy in Education, p. 30.

Little is done to assess students' thinking during learning so we can help them learn better. Nor do we collect and aggregate student-learning data in ways that make the information valuable to and accessible by educators, schools, districts, states, and the Nation to support continuous improvement and innovation. We are not using the full flexibility and power of technology to design, develop, and validate new assessment materials and processes for both formative and summative uses." With better technology, as well as better regulations and standards concerning data, the end-of-year assessments required for schools to secure Federal funding can be made more meaningful and at lower cost. We elaborate on this topic in Chapter 6.

Need for an Ongoing Process. The state-led movement to establish standards and assessments in mathematics and science represents an historic opportunity. This activity, however, must be a long-term process, not a one-time event. The changes envisioned by new standards and assessments require a longterm investment by states and by the Federal Government in the nation's educational infrastructure.

CONCLUSION: A CALL TO THE STEM COMMUNITY AND TO THE FEDERAL GOVERNMENT

Due to efforts over the past quarter-century at many levels and across the political spectrum, the United States has the opportunity to come together around shared, high standards for both mathematics and science education. It is important to seize this historic moment.

PCAST strongly supports the current state-led movement to develop shared, high standards in both mathematics and science. These standards should reflect the mix of factual knowledge, conceptual understanding, procedural skills, and habits of thought described in recent studies by the National Research Council.

PCAST calls on the science, technology, engineering, and mathematics community in industry and academia to actively support the state-led shared standards effort. **It will be important for the broader STEM community in higher education and industry across the Nation to show their support and speak to the value of shared standards and assessments. Moreover, it is important that STEM professionals and organizations look beyond their individual objectives for mathematics and science standards and focus on the greater common goal the Nation has to complete the standards and ensure their widespread use.**

As state-led consortia develop, adopt, and/or align their standards to shared standards in mathematics and science, ***the Federal Government should provide resources for the development of aligned assessments and for professional development.***

Finally, the creation of high-quality shared standards and assessments is not a one-time effort but will require a sustained focus over a decade. Any set of standards will need to be continually reviewed and improved based on research on use and effectiveness. We urge the states, the Federal Government, the STEM community, and the Nation to develop mechanisms to ensure that this will occur and to support the continual, high-quality research that will be needed to inform this work.

We urge the states to sustain their commitment to establishing and implementing shared standards for mathematics and adopting new high-quality assessments. We urge them to do the same for new science standards once they are released. They need to put into place formal mechanisms that will maintain an intense focus for at least a decade. We also urge the Federal Government to continue its strong support for the state-led movement. It should use its policies and resources to ensure continued progress by providing powerful incentives in the form of financial assistance to states that adopt standards and assessments tightly aligned with the shared standards. We also urge the broader engagement of the Nation in this effort, including STEM-related corporations, colleges and universities, and other groups.

Federal Support for the State-Led Standards Movement

The Federal Government can provide valuable support for the state-led movement to develop and adopt shared standards and for efforts to develop assessments aligned with those standards. The process of adopting new standards can be expensive for states and school districts. States face the expense of developing, acquiring, administering, and improving over time the assessments that measure student learning of STEM skills and knowledge. Even the relatively simple high-stakes annual assessments required under No Child Left Behind cost states more than $800 million per year.[108] Tests that more deeply measure STEM learning and knowledge will be more expensive, as described earlier in this chapter.

Here, the Federal Government can support states and nonprofits to develop and continue to improve assessments. These investments can be more efficient if targeted at consortia of states that develop assessments to meet shared standards rather than on a state-by-state basis. Consortia also make sense because development of valid and reliable assessments requires considerable expertise, which is limited and cannot effectively prepare high-quality tests for many different states. Given the expenses of developing high-quality assessments, financial support can provide a powerful incentive for states to adopt shared standards more quickly.

Beyond developing standards and assessments, teachers need to know how to use the standards and the assessments. They also need tools that enable them to teach to the new levels of performance those standards will describe. The Federal Government has a history of investing between $3,500 and $4,800 per teacher on science and mathematics teacher professional development tied to the NSF's systemic initiatives – but this does not mean that this is the level of funding required for effective or cost effective professional development. We estimate that at least this amount, and likely much more, will be needed to support teachers to embrace new standards.[109] This can be a disincentive for states, schools, and school districts to adopt the standards, especially in a difficult economic climate. The Department of Education can help provide the financial support needed for professional development that helps teachers and school leaders adapt their curriculum and teaching methods to implement shared science and mathematics standards. Cost estimates should consider a combination of face-to-face workshops and online professional development, beginning with support from external providers and over time moving to the professional development that occurs when prepared teachers work cooperatively within a school. Costs should include a rigorous research component to enable continuous learning and feedback about what works, for whom, and why. We address the challenge of providing cost-effective professional development in STEM in Chapter 5, Recommendation 5-2.

The Federal Government already provides about $400 million per year to support state testing. In addition, the Federal Government has committed $350 million to develop new assessments for college- and career-ready standards and high-school end-of-courses exams through state consortia. Considerable additional funds will be required over the next five years to develop new assessments in science, as new science standards are developed and adopted by the states, and to ensure ongoing improvement of both math and science assessments.[110] These investments will be well worth the price because they will make it feasible for states to adopt new standards with assessments that strongly support the new learning goals, provide better data to improve STEM education initiatives, and ultimately make assessment development more efficient and cost-effective on a national level. Finally, the Federal Government should support states to give benchmarking exams that allow comparisons of U.S. students to students around the world.

RECOMMENDATION 4-1. STANDARDS AND ASSESSMENTS

We urge the states to continue their historic efforts to adopt shared standards in both mathematics and science. Math standards were released in June 2010, and we applaud the many states that have already adopted them and urge other states to do so. Science efforts are at an earlier stage; we underscore the importance of this effort and urge that standards be developed by 2011 and adopted by states soon thereafter.

We recommend that the administration support this state-led effort by providing financial support and technical assistance to make it feasible for states to adopt such standards.

1. **Support for Standards. The Department of Education should provide support for rigorous, high- quality professional development tightly aligned with the needs of teachers to implement curricula that meet shared standards. It should do so by providing funding to states and to nonprofit and for-profit organizations that provide technical assistance to states and districts for professional development. It also should provide support for the development of high-quality online resources for professional development.**
2. **New Assessments. The Department of Education should expand its support to consortia of states to pay for the cost of development, evaluation, administration, and ongoing improvement of assessments tightly aligned to the standards. These assessments should foster active, inquiry-based learning together with mastery of critical concepts and facts. They should include both periodic cumulative assessments and end-of-course assessments for selected courses in high schools.**
3. **Ongoing Research. The National Science Foundation and other foundations and research organizations should support research on the transition to shared standards, evaluating their success, the challenges faced, and ways to improve them over time.**

Concerning support for professional development for new standards, we estimate a cost of about $450 million per year for five years for the Nation's roughly 450,000 public school mathematics and science teachers, at a cost of approximately $1,000 per teacher per year. In the Department of Education's request for reauthorization of ESEA, $300 million is requested in FY11 for effective teaching in mathematics and science that can be used toward this end.

However, we estimate that an additional $150 million per year will likely be needed from other Federal, state, or private sources for a period of 5 years.

Concerning new assessments, we see two important roles for the Federal Government: first, ongoing support for research and development of new high-quality math and science assessments, and second, Federal support to assist states with implementation and annual administration of these assessments.

The Federal Government has currently allocated $330 million for the development of new mathematics and English language arts assessments, which we estimate should provide adequate support for new, high-quality math assessments to emerge. Given that the expense of implementing and administering these new assessments could be prohibitive for states, we also recommend that the Federal government provide $250 million per year in Federal funding to States to administer new mathematics assessments, which may require supplementing, or specifically targeting toward new assessments, the Federal Government's commitment of $400 million in annual support to states for assessments in all subject areas.

For science, once the new standards are released, we estimate that the Federal Government should provide $10 million to support the development of a new science assessments framework – a research- informed planning process to lay out the key goals of and logic models for new science assessments. This should be followed by at least $160 million in Federal support for the development of new science assessments by consortia – similar to the current competition underway for mathematics assessments. To support states to implement and new science assessments after they are developed, we estimate that $250 million to $300 million per year will be required in annual Federal support to the states for administering science assessments, which may require supplementing, or specifically targeting toward new assessments, the Federal Government's commitment of $400 million in annual support to states for assessments in all subject areas.

In supporting the states to determine the right combination of assessments that will best reflect what students need to learn to be ready for careers and college, the Federal Government might also consider supporting development and state implementation of end-of-course assessments in high school for at least one course in mathematics (algebra or above) and in one in science (biology, chemistry, or physics).

Producing high-quality assessments for science will be an especially difficult and demanding task. To recruit some of the Nation's best scientists, teachers, and assessment experts to help in this critical effort, we believe that the NSF should fund the National Academies to produce, on a rapid time scale of less than a year, a *Framework for the Assessment of Science Education*, based on the *Framework for Science Education* when it is published. The assessment framework should address recommendations for both formative and periodic cumulative assessments.

Recommendation 4-2. Benchmarking

The Department of Education should provide financial incentives to states to participate in currently available, high-quality national and international assessments in order to monitor progress relative to national and international benchmarks. It should continue its support for NAEP testing and should provide support for PISA and TIMSS.

A first step toward implementing this recommendation could be to link NAEP to international tests to allow NAEP results to be used in international benchmarking. In the long term, direct participation by the states in PISA and TIMSS would be valuable. We estimate that in addition to the current Federal support for NAEP of approximately $130 million per year, approximately $85 million in Federal funding every four years will be required to enable all states to minimally participate in TIMSS and NAEP.

V. Teachers

I can truly say he molded me.

– Angela Fajardo, a teacher and a former student of the legendary Los Angeles mathematics teacher Jaime Escalante, in an interview with the Los Angeles Times.[111] Escalante, who led poor Hispanic students at Garfield High School to achieve high scores on AP Calculus exams and pursue career paths in mathematics, was brought to fame in the movie *Stand and Deliver*. Escalante passed away during the preparation of this report.

Chapter Summary

Teachers are the single most important factor in the K-12 education system, and they are crucial to the strategy of preparing and inspiring students in STEM. Great STEM teachers have at least two attributes: deep content knowledge in STEM, and strong pedagogical skills for teaching their students STEM. These two attributes allow great teachers to help students achieve deep understandings of STEM that they can use in their lives and careers. These attributes also enable teachers to excite students about the dynamic nature of STEM fields, motivating them for lifelong study. Too few of these teachers are in the Nation's classrooms, in part because of a lack of professional respect, the inconsistency of teacher preparation programs, and the salary disparity of teaching relative to other STEM fields. The Federal Government should help recruit, prepare, and support at least 100,000 new STEM teachers with these attributes at the middle school and high school levels over the next decade. It also should support the professional development of all teachers to help them achieve deep STEM content knowledge and mastery of STEM pedagogy. And the Federal Government should recognize and reward the best STEM teachers nationwide through the creation of a STEM Master Teachers Corps.

Introduction

Anyone who has set foot in a classroom knows that teachers make a huge difference in the lives of their students. While not everyone can recall the influence of a legendary teacher like Jaime Escalante, many people have stories about the crucial role that a teacher played in sparking their passion for a subject, teaching them a lifelong skill, or helping them surmount an obstacle. Indeed, most people who work in STEM fields or who simply have an interest in

STEM can point to teachers who excited about them about the beauty and power of mathematics, the wonders of science, or the power of technology – and who helped them learn that they could gain mastery of these subjects. Sadly, though, many adults also point to experiences in school that convinced them that STEM subjects were inherently boring, cryptic, or beyond their grasp.

Teacher quality clearly matters in student learning and success. And great teachers can motivate students to devote themselves to a subject and overcome obstacles in the learning process, as well as inspire students to pursue lives of discovery and invention. The critical questions are: What attributes characterize great STEM teachers? How can we best recruit and train great STEM teachers? And how can we support them and retain them? In this chapter, we discuss these issues and make specific policy recommendations.

The Challenge of Understanding Teacher Impact

Teachers are arguably the single most important component of education that can be influenced by policy. We know that teachers are not all the same, and that great teachers expand how much and improve how well students learn. Yet, there are vast gaps in our knowledge about how to produce great teachers and retain them. Many studies are not directly useful in policymaking, owing to inadequate or inconsistent research design or data collection. Based on what research has emerged over the past two decades, however, we know the most about how students learn, the next most about the characteristics of effective teaching, and the least about characteristics of programs that are effective in producing effective teachers – let alone great teachers. We briefly elaborate on these points below.

That teachers have a large, measurable impact on student learning and achievement is clear. A National Research Council panel recently completed a five-year study, entitled *Preparing Teachers: Building Evidence for Sound Policy*, which examines what is known about effective teacher preparation in mathematics, science, and reading. The report notes that there is broad agreement and clear evidence that "teachers have enormously important effects on students' learning, and that the quality of teaching explains a meaningful proportion of the variation in achievement among children."[112] Various studies have demonstrated this impact in different ways. For example, researchers have shown that teachers who see gains in student achievement in their classes tend to see them repeatedly over time.[113] Also, students taught by a series of teachers deemed exemplary for previously raising test scores learn and achieve more.[114 115] Research indicates that a teacher who has consistently raised student achievement can make a greater difference in student outcomes than more costly interventions such as reducing class size.[116] The National Mathematics Advisory Panel has pointed to evidence showing that variation in teacher quality could account for a substantial fraction of the total variation (as much as 12 to 14 percent) in mathematics learning by elementary students in a given school year.[117]

Despite the broad agreement on the importance of teachers, the NRC report concluded that there is little solid research about precisely *how* and *why* teachers influence student outcomes and about *how* teacher preparation programs should be designed to train great teachers. How much students learn depends on many factors, both internal and external to the education system, and it is difficult to carry out research that can definitively distinguish causation from correlation to identify the effects of teaching. Many studies about connections between teachers and student outcomes have not yielded truly meaningful conclusions. (We address the need for improved education implementation research in Chapter 3.)

We concur with the National Research Council committee that more work is desirable to rigorously evaluate what teachers contribute to student learning and achievement, what teacher qualities are most important, and which programs work best to prepare teachers. To inform and guide policy, we also need better measures of student learning than standardized test scores and finer grain measures of the attributes of teachers and teach preparation program.[118] Truly rigorous research on teacher preparation would require educating and preparing many thousands of new teachers through a variety of different teacher preparation programs and then measuring the impact that they have on student achievement. And truly rigorous research on teacher retention will require testing a variety of approaches to assess their affect on the long-term retention of the best teachers.

Realistically, it will take 10 to 20 years before definitive research is available that defines the best ways to recruit, prepare, and retain the great STEM teachers that the Nation needs to prepare and inspire students for the 21st century. Yet the Nation cannot wait 10 to 20 years to act. We are already expending considerable funding and effort to train tens of thousands of teachers every year. It is possible to act wisely now based on a combination of the best available evidence, professional judgment, and common sense. At the same time, we should design our actions to collect additional evidence along the way. In short, we must act and learn at the same time.

Finally, it might be argued that we should not focus on "inputs" (understanding the qualities that make a great teachers, and funding programs that are most likely to produce such teachers), but rather should primarily exclusively on "outputs" (measuring the impacts of teachers in classrooms, in response to which prospective teachers and teacher training programs will adjust). We disagree, believing that it is important to focus on *both* inputs and outputs. As with the research issues discussed above, it will likely take a decade or more for the "market" for training and hiring of new teachers to respond fully to output measures. The Nation will be well served by investing in encouraging the best teacher training programs.

Attributes of a Great STEM Teacher

Despite the gaps in research, the available evidence, professional judgment, and common sense converge on the idea that great STEM teachers have at least two important attributes: (1) They have deep content knowledge, and (2) they have strong pedagogical training specific to STEM. These two attributes help great teachers both prepare and inspire students in STEM.

1. **Deep content knowledge**. Great STEM teachers understand their subject matter deeply enough that they can explain concepts and procedures from multiple perspectives, thus guiding students to explore on their own. They stay current with the evolving body of knowledge in their fields, so that they can draw on it to make their subjects come alive. They know enough about their subject matter, and its relationship to issues in the real world and current events, to make STEM relevant. Teachers require enough content knowledge to deal with questions from inquisitive

students and, in turn, pose challenging questions to their students. Excellent STEM-trained teachers know enough that they can turn questions (Why can't you divide by zero, or take the square root of a negative number?) into teachable moments, rather than simply saying, "Because it's a rule." Teachers with this kind of knowledge can ignite student interest in STEM, inspiring them for lifelong study. They also encourage students to question assumptions, rather than accepting what they are told as a given. They cultivate within students the capacity to pose probing questions and to figure out methods of answering those questions, rather than simply teaching them to answer predictable questions. In this way, these teachers allow students to practice and develop the essential capacity to reason the way that scientists, engineers, computer scientists, and mathematicians do.

There is little doubt, in the accounts of current STEM professionals and students of great STEM teachers, that deeply knowledgeable teachers can inspire and motivate students to study STEM. The precise extent to which such teachers better prepare students by improving their achievement and learning is not yet clear, but several sources of evidence point to the fact that deeper STEM knowledge makes for better STEM teachers. Researchers have shown, for example, that higher scores for elementary school teachers on tests of mathematical knowledge for teaching[119] predict higher gains in student achievement.[120] The recent National Research Council report on teachers concludes that in mathematics in particular "there is strong reason to believe that teachers' knowledge and skills make a difference in their practice."[121]

2. **Mastery of pedagogy.** Content knowledge alone is not enough. Great teachers must be able to call on a repertoire of methods that help them manage their classrooms and illuminate STEM topics. Importantly, they must have a nuanced understanding of how learners approach STEM subjects. These teachers know the natural misconceptions that arise from intuition and can challenge misconceptions in ways that help students relinquish them based on a real understanding, not memorization of a fact or a phrase. They can guide students in scientific inquiry, the design of experiments, and making sense of data. They also know how to motivate and excite students to learn topics in STEM. Such tactics can be taught in teacher preparation, induction, and professional development programs that give STEM teachers specific tools and approaches they can use in their classrooms. Teachers need to hone these skills and adapt them to the curriculum and their students. This requires that they have ongoing support and peers with whom they can share best practices.

   There is a broad consensus among STEM education professionals that pedagogical mastery specific to STEM is an essential quality of great STEM teachers.[122] Evidence strongly points to the kind of pedagogy that is most effective in STEM education: cooperative, collaborative, active, and inquiry-based methods increase learning and retention of information and higher order thinking skills.[123] It is important that STEM teachers both prepare and inspire their students, engaging them in STEM as much as they instruct them.[124] Reports on STEM learning undertaken by the National Research Council have noted the importance of teachers acquiring a command of the strategies needed to illuminate STEM for learners, and of teachers knowing the variety of ways in which learners develop STEM knowledge and skills.[125]

We should underscore that these two attributes do not *guarantee* great STEM teachers. Some teachers may fail to be effective despite these characteristics. Ideally, we should assess teacher effectiveness, on an individual basis, based on student achievement and interest in STEM. Nevertheless, these two attributes are valuable proxies in considering the types of teacher preparation programs that can produce great teachers.

Supply and Demand for STEM Teachers

What is the ongoing national demand for STEM teachers in public schools? Currently, there are an estimated 477,000 teachers whose main teaching assignment is mathematics or science in K-12 public schools, of which about 426,000 teach in middle and high schools.[126] Each school year, there is considerable turnover in the STEM teaching force. If recent trends continue, about 25,000 mathematics and science teachers can be expected to leave the profession annually. Less than a third of them retire, while nearly two-thirds cite job dissatisfaction as their reason for leaving.[127] Within the first five years of teaching, more than 40 percent[128] of teachers decide they no longer want to teach, many due to a lack of professional support. Although many teachers re-enter the workforce and change locations and schools, which contributes to this turnover, for planning purposes we will assume an ongoing average need for 25,000 STEM teachers – although the need could increase if we succeed in attracting more students to STEM fields.

The overall supply of mathematics and science teachers nationally has been rising to meet total demand, but there are local imbalances on a regular basis, with many schools struggling to fill openings in STEM subjects with qualified teachers. In particular, schools in high-poverty communities often do not have access to knowledgeable teachers. This occurs, in part, because STEM teachers migrate within the profession to better paying jobs at better-funded schools.[129]

However, the issue is not just the quantity of STEM teachers available: there is a crucial issue of quality. Many existing STEM teachers and teachers in training programs today lack the attributes associated with great STEM teachers.

The data are clearest for mathematics, although observations suggest that it is likely true for science as well. The recent NRC report on teacher preparation concluded that "many, perhaps most, mathematics teachers lack the level of preparation in mathematics and teaching that the professional community deems adequate to teach mathematics," and that too many middle and high school teachers teach mathematics out of their field.[130] The report also concluded that too little information is known about the preparation of science teachers, citing a 2003 survey showing that 28 percent of the U.S. public school teachers who are teaching science in grades 7-12 lack a minor or major in the sciences or science education.[131] Evidence also suggests that the problem is more acute in schools in high-poverty communities; a 2008 study showed that 40 percent of mathematics classes in high-poverty secondary schools were taught by out-of-field teachers, whereas 83 percent of classes are taught by teachers with mathematics or mathematics education degrees in schools that serve the fewest low-income students.[132]

It should be noted that the available data tend to focus on whether teachers meet a bare minimum standard of having a major or minor in their fields. There are far fewer data about the proportion of teachers who have deep content knowledge of their field and a mastery of how to teach learners of STEM, which constitutes a higher standard.

The considerable annual demand for new teachers offers a tremendous opportunity to improve the quality of the STEM teacher workforce through improved recruitment, preparation, professional development, and retention.

Preparing Great STEM Teachers

STEM teachers in U.S. public schools are currently trained in a variety of ways. These "pathways" generally fall into two categories: traditional programs based in colleges or universities that lead to a BA or MA and teaching certification, and alternative programs to certify teachers, which vary widely in their characteristics.[133] Determining exactly how many teachers are prepared through each of these pathways is difficult given a general lack of data. The distinction between the two pathways has also become increasingly blurry. Importantly, the programs *within* the pathways vary widely in their requirements for developing the attributes of great STEM teachers.[134]

Even among the teachers who take the university or college pathway to becoming educators, many do not acquire strong majors in STEM subjects paired with pedagogical training. Few teacher preparation programs put an emphasis these two critical attributes of great STEM teachers.[135] The problem is particularly acute in elementary teacher preparation programs but is also serious for middle and high schools.[136 137]Although data are scarce on the precise level of preparation of the Nation's teachers in the critical STEM content knowledge and pedagogical skills required to excel in the classroom, knowledgeable observers agree that far too few teachers have been adequately prepared and supported to develop these two attributes. What evidence is available suggests that teachers seldom are prepared by both the education schools and STEM departments within their colleges, universities, or certification programs – and that they seldom have a strong major in STEM. For example, during the 2007-08 school year, 56 percent of K-12 public school mathematics and science teachers held undergraduate and/or graduate degrees in mathematics, mathematics education, science, or science education. (About 60% held such degrees at the middle and high school levels.) Yet only about 25 percent of science and mathematics teachers had an undergraduate or graduate degree conferred by a STEM department or school.[138] Even among the small fraction of teachers trained deeply in both education practice and STEM fields, we know little about the quality of instruction they received in STEM content or STEM pedagogy.

Ideally, we would know the characteristics of preparation programs that best cultivate the kinds of teachers who prepare and inspire K-12 students in STEM. Available research does not show significant differences between the major teacher pathways with respect to teachers' knowledge and effectiveness, but this reflects the very coarse nature of the studies to date. The National Mathematics Advisory Panel concluded that not enough is known about teacher education to "draw conclusions about the features of professional training that have effects on teachers' knowledge, their instructional practice, or their students' achievement." [139] This dearth of information underscores the need to collect thorough, consistent data on teacher preparation and to design studies on teacher programs and student outcomes that allow policymakers to identify the characteristics of truly effective programs. (We address this need in chapter 3.)

Individual programs that provide strong pedagogical preparation in STEM, by preparing teachers to convey concepts and lead inquiry in the classroom, and deep content expertise via rigorous majors in STEM disciplines have provided examples of success. (One example of such a program, UTeach, is described in Box 5-1.) What such examples suggest is that effective teacher preparation programs will achieve multiple goals. They will attract college students with strong backgrounds or aptitudes for STEM subjects and an interest in teaching, provide them with rigorous extended pedagogical training based in the classroom, and prepare them to respond to ever-changing STEM knowledge and the dynamics of their schools.

Box 5-1. UTeach

UTeach, a teacher preparation program launched in 1997 at the University of Texas-Austin and now in 20 universities, recruits undergraduate mathematics and science majors to pursue teaching careers and rigorously trains these future teachers in both STEM content and STEM-specific pedagogy. The program draws on faculty from education and STEM departments, bridging what is typically a gulf between education schools and schools of science, mathematics, and engineering. UTeach allows students to earn both a degree in a STEM subject and their teaching certification within a four-year period. It also provides intensive in-classroom training and supports new teachers during their first formative years of teaching in classrooms, including providing mentors who visit new teachers' classrooms and coach them on content- specific pedagogy.

Because UTeach spans departmental boundaries at universities, it can effectively recruit mathematics, science, and computer science students into teaching, even if they are not initially enrolled in education courses. Science and mathematics faculty, who are often role models to high-achieving college students, actively encourage STEM teaching as a career route. The program attracts particularly strong students; at nearly every university site the average SAT or ACT scores of UTeach participants is higher than the average for that university's undergraduate population as a whole.

Five years after they begin teaching, 82 percent of UTeach graduates are still teaching, compared with national retention rates of less than 60 percent. And 45 percent of the teachers who graduate from UTeach programs have chosen to teach in high-need schools, in part due to the program's emphasis on public service.

UTeach has been replicated by 21 other universities around the country, who together with the flagship UTeach program plan to recruit and prepare more than 4,500 mathematics and science teachers by 2015, and 7,000 mathematics and science teachers by 2018. The cost per graduate of the program is about $8,000, with the costs split roughly equally between the state universities and private sources. UTeach's leadership projects that $80 million would be required to produce 10,000 STEM teacher graduates per year at 200 UTeach programs around the county. That figure is based upon a cost of $2.5 million for each new university UTeach program.

RECOMMENDATION 5-1. 100,000 STEM TEACHERS FOR MIDDLE AND HIGH SCHOOLS

At the middle and high school levels, the Federal Government should set a goal of ensuring the recruitment, preparation, and induction support of at least 100,000 new STEM teachers over the next decade from programs that are (i) designed to produce teachers who have strong majors in STEM fields and strong content-specific pedagogical preparation – including teachers from nontraditional backgrounds who help diversify the STEM teaching force – and (ii) capable of measuring both the student achievement and the retention of the teachers they produce.

The target number of teachers per year would average 10,000, but it may require some time to ramp up to produce 100,000 high-quality STEM teachers over a decade.

Toward this end, the Department of Education should systematically identify programs with the characteristics described above and should provide financial support for the best programs to scale up to significant levels. These programs should regularly assess their success with respect to student achievement and teacher retention relative to appropriate comparison groups, so that we continually learn how best to prepare great STEM teachers.

The Department should gather data and report annually on progress toward this goal.

Teachers require early career support and thrive best in schools where they can learn from one another and where they are led by strong principals who are committed to high-quality STEM instruction.

Without substantial and immediate knowledge of the outcomes of a broad range of teacher preparation programs, it makes sense to expand programs that have a high likelihood of preparing great teachers, while simultaneously improving the research enterprise so that data drives the evaluation and selection of such programs in the future.

The Nation should therefore invest in teacher preparation programs that provide strong content knowledge preparation, pedagogical training, and induction support to teachers, and that collect data about teacher retention and teachers' impact on student learning and achievement. In making such investments, it will be important to focus particularly on programs that are *scalable*, because they will have the greatest impact in terms of the number of teachers produced and the greatest opportunity for learning about successful elements of programs.

In identifying programs that *merit* scale-up, it will be helpful to look for the following features: close collaboration between K-12 schools and the teacher preparation program, including the involvement of practicing teachers and administrators in teacher preparation; collaboration between faculty in STEM departments and schools of education, enabling new teachers to learn deep content expertise as well as the teaching and learning of subject matter; on-going support that recognizes learning-to-teach as a process that takes years; and facilitation of supportive working conditions, including good instructional materials, clear guidance, strong administrative leadership in schools and districts, and networks linking teachers who are committed to ambitious STEM instruction.

It is also important that stakeholders work to identify the characteristics of teacher preparation programs that *allow* successful replication and scale-up. These features may include cost-effectiveness, a structure and organization that are transferable to various institutions and locations, a model for leadership and implementation that can be taught to others, and materials and tools that can be readily adapted and disseminated.

Finally, we note that some may wonder why the Nation should invest in new teacher training at a time when the media carries stories about teacher layoffs or hiring freezes. We believe that the current labor market for teachers represents a response to temporary economic conditions. Over the longer term, the overall demand for new teachers will surely resurge. Moreover, science and mathematics teachers remain in demand even in times of high unemployment and cutbacks.

We estimate that approximately $100-150 million per year will be required for 10 years, assuming a per- teacher cost of at least $8,000 and at least $2,000 per teacher-graduate in administrative costs. Pending reauthorizations of Federal legislation, if fully funded, could provide significant resources for this recommendation. In the reauthorization of ESEA, the proposed Teacher and Leader Pathways Program ($405 million annually) and Effective Teachers and Leaders Program ($2.5 billion annually) could provide a substantial portion of the necessary funding. Through reauthorization of America COMPETES, full funding of the Noyce Fellowship Program at the NSF (at $140.5 million for section 7002 and 7030) could allow for funds to be channeled to implementation of this recommendation.

While this recommendation pertains to middle and high school teachers, it is also important to produce great STEM teachers at the elementary level. We urge the Department of Education to develop a plan to identify and bring to scale proven methods to improve the expertise of elementary teachers, whether through specialists, coaches, lead teachers in each school, or other methods.

In addition to recruiting and preparing new STEM teachers, it is important for current STEM teachers to be able to improve their knowledge and skills. The Federal Government has an important role to play in supporting research about effective professional development and about developing new types of professional development programs. While there are many professional development programs for STEM, they are a mixed bag in terms of quality and impact. There is a need to identify the types of programs that truly benefit teacher skills and student achievement. Available studies indicate that effective professional development is tied explicitly to the curriculum taught to students, preparing teachers to use materials effectively based on other teachers' successes and to help address student questions. Professional development where teachers evaluate videos of their own teaching with a facilitator and in which teachers discuss how to evaluate and understand student progress based on assessments have been successful. To make professional development more affordable for teachers, schools, and school districts, there is a need to develop programs that are both effective and cost-efficient. Courses with online components may prove particularly worthy of investigation.

High-caliber professional development and support are critical for helping many of the Nation's STEM teachers perform at their best. In addition, elementary school teachers, who tend to be generalists rather than STEM specialists, require special consideration in professional development programs.

RECOMMENDATION 5-2. PROFESSIONAL DEVELOPMENT FOR EXISTING STEM TEACHERS

The Federal Government should vigorously support research to identify and develop high-quality, cost- effective STEM teacher professional development programs and should invest in the dissemination of those programs.

We believe that much of the research and development needed to identify and create these professional development programs can be done with existing resources at the NSF and the Department of Education. Additional resources will be required for dissemination of such professional development programs, as described in recommendation 4-1.

Rewarding and Professionalizing Great STEM Teaching

Excellence in STEM teaching requires a focus not only on preparation but also on recruitment and retention. Two major factors must be addressed: professional respect and salaries.

1. **Professional Recognition and Respect.** STEM teaching lacks the professional stature and prestige it deserves given the influence that STEM teachers will have on the future of our country. STEM teachers are underappreciated and under-recognized. Many lack role models, mentors, and networks of support. STEM teachers lack sufficient opportunities for professional growth that allow them to do what they love: teach.

   Science and mathematics teachers who leave the profession frequently cite job dissatisfaction as their reason for departure. After the 2003-04 school year, survey results show that 14,000 of the 26,000 mathematics and science teachers who stopped teaching were not satisfied with their jobs.[140] When STEM teachers leave their schools for other schools, they also frequently cite their professional frustrations as a reason. What lies beneath such dissatisfaction is complex, but when STEM teachers leave their schools they often point to a lack of useful professional development and to problems they have with classroom management. In addition, many STEM teachers are frustrated by unresponsive school, district, and state management systems. Unless we give our best teachers access to decision-makers and a voice in the many policy decisions that affect their lives, we will continue to make poor use of the talented teachers in our schools, and large numbers of them will leave the profession for jobs that are more responsive to their needs.

   In short, we need to treat STEM teachers – and, indeed, all teachers – as professionals if we hope to attract and keep great teachers in our schools. Teachers need access to relevant professional support and to peers who have grappled with similar problems, and they need to feel that their work is respected and recognized. They need opportunities to serve as leaders among their peers and in their profession. In turn, we need to expect of teachers the accreditation and performance of professionals.

The program Math for America includes programs to provide excellent mathematics teachers with respect and rewards (see Box 5-2).

2. **Salary.** Salaries for public school teachers are set by state and local authorities, and the salaries vary across jurisdictions. Nonetheless, it is fair to say that, in purely financial terms, it does not pay to be a STEM teacher in the United States.

BOX 5-2. MATH FOR AMERICA

Math for America is a relatively new program that began in New York in 2004 and has recently expanded to Boston, Los Angeles, San Diego, and Washington, DC. Math for America uses two approaches to improving the quality of mathematics teaching in the cities it serves.

The first is a fellowship program that recruits and trains skilled mathematicians to serve as K-12 teachers in public schools by paying their tuition to receive mathematics education master's degrees and supplementing their salaries for their first four years of teaching. Math for America recruits new teachers primarily through outreach to college and university campuses, in career fairs, and in meetings with mathematics departments and faculty. The program is highly selective, seeking out potential teachers who were above average achievers as college students and who have a high aptitude for mathematics. (The 2010 class of fellows had an average mathematics SAT score of 721 and an average GPA in college of 3.47.) A proportion of these new mathematics teachers are career-changers who heard about the program and decided to leave their jobs in finance or engineering to pursue teaching. Ninety percent of the new teachers who participate in the teacher training fellowship program stay in the profession through their fifth year, compared with rates of less than 50 percent in similar districts around the country.

The second approach is a program that confers prestige and salary supplements to exemplary mathematics teachers in public schools. The organization has awarded fellowships to master teachers in New York City public schools, providing annual stipends, professional development, and leadership opportunities to exemplary teachers to encourage them to be more involved in their schools, districts, and profession. The master teachers say the program heightens the importance they place on their jobs, because they serve as mentors and leaders to other teachers and earn recognition and salary in exchange.

Math for America expends an estimated $25,000 per year for each new teacher trainee and between $15,000 and $20,000 per year for each master teacher awarded a fellowship. Both programs provide mentoring, education, and professional support specific to mathematics instruction. Networking opportunities help new and experienced mathematics teachers excel. Math for America aims to create a sense of solidarity, professionalism, and support for its teachers. It fosters a community by holding workshops three times a week for its fellows on mathematics teaching methods and content, conferences, and social events. It assigns mentors to new teachers and hosts a virtual network through which new and expert teachers ask each other for advice and share knowledge about mathematics teaching.

> College graduates with comparable educational backgrounds to high-quality STEM teachers often pursue careers in which they have greater earning potential. The median salary offer for recent college graduates going into elementary teaching is $30,000; the median salary for new secondary teachers is $36,000.[141]

By contrast, the median salary offered to recent college graduates in certain STEM-related fields, including physics, computer science, accounting, and engineering, is more than $60,000.[142]

> Even if one accounts for the fact that teachers work for a nine-month period and could seek summer employment to obtain some additional salary, the gap is substantial. Moreover, salaries in STEM teaching have not kept pace with other STEM professions. The median salary for high school science and mathematics teachers in the United States grew by 8 percent adjusted for inflation between 1993 and 2003, while the salary gains in STEM professions that represent alternative career trajectories for individuals with STEM degrees grew by 21 to 29 percent.[143]
> In international comparisons, teacher salaries in the United States lag behind most developed countries, even though teachers here work more hours on average.[144] Relative to per capita GDP, the U.S. ranks in the bottom third of OECD countries in terms of teacher salary (see Box 5-3). U.S. salaries even lag in absolute terms behind countries like Germany, Japan, South Korea, and Switzerland.[145]

Box 5-3. Salaries of Secondary* Teachers Relative to GDP in OECD Nations

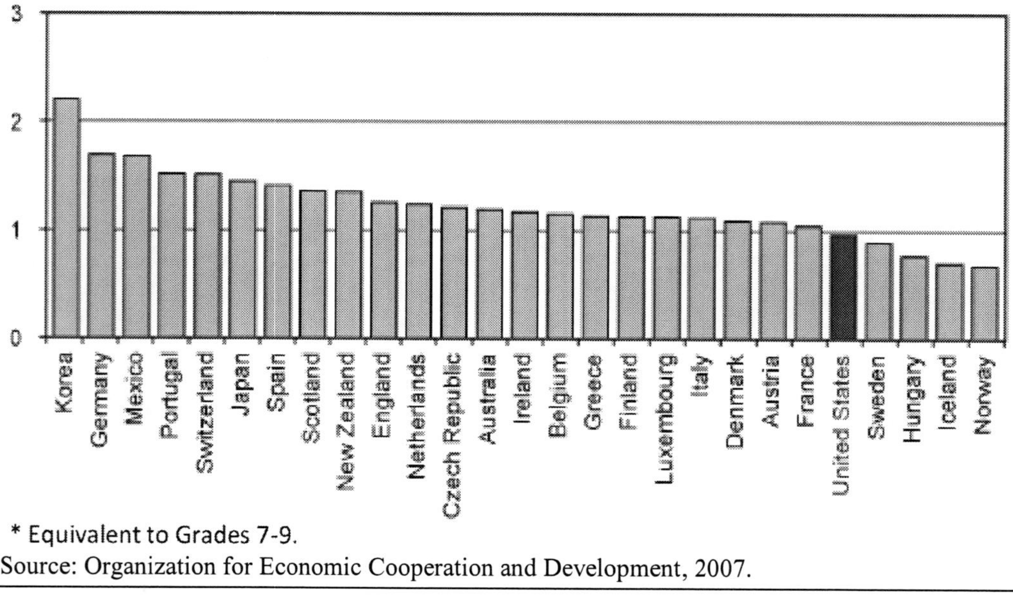

* Equivalent to Grades 7-9.
Source: Organization for Economic Cooperation and Development, 2007.

Ideally, school systems across the Nation would significantly improve salaries and working conditions for all STEM teachers across the board. For a variety of reasons, this is not economically feasible, at least in the short term. Nonetheless, we believe that the Federal Government can make a significant impact on the profession by recognizing and rewarding a significant fraction of the very best STEM teachers in the United States.

The Federal Government administers some programs to recognize excellence in teaching, but they touch only a minuscule fraction of the STEM teaching force. The Presidential Awards for Excellence in Mathematics and Science Teaching Program provides a one-time award of $10,000 to STEM teachers, but the annual number of awardees is only about 100, corresponding to far less than 0.1 percent of STEM teachers in the United States. The Robert Noyce Scholarships and the Leonore Annenberg Teaching Fellowships together award scholarships to about 950 aspiring and new teachers each year who teach mathematics and science in high needs schools.[146] These programs are commendable and have an impact on the individuals recognized. But the programs are far too limited to have a significant impact on the profession overall.[147]

STEM Master Teachers Corps

To attract and retain great STEM teachers, we must significantly and visibly reward excellence in STEM teaching, signal the importance of the profession, and elevate the level of STEM teachers by setting a new high bar for excellence. This requires recognizing a substantial number of STEM teachers nationwide and creating a network among them that can drive progress and aspiration in the profession.

Toward that end, PCAST believes the Nation should create a STEM Master Teachers Corps that recognizes the best STEM teachers in the Nation. Such an organization would reward excellence, raise the status of the profession, and attract and retain the best teachers. We believe that a STEM Master Teachers Corps could send a powerful signal.[148]

A precise plan for such a Corps will require additional deliberation, but we sketch key issues. Because there are various open questions about implementation, we recommend that the administration convene a focused working group to address the issues and design a pilot program.

1. **Selection.** Teachers should ideally be selected based directly on their ability to prepare and inspire students. This would require assessing *impact on students* in a very broad sense – not merely their ability to elicit improvements in standardized test scores, but also to spark interest and actually induce students to reach farther. Realistically, the criteria would need to include objective and subjective measures of teacher knowledge and skill and student achievement. Careful design of these criteria is essential to ensure that they do not disproportionately exclude teachers in underperforming schools. Rather, the selection process should give particular consideration to teachers whose service in high-needs schools further demonstrates their excellence as STEM teachers. Moreover, efforts should be made to represent a range of grades and subject areas within STEM. Independent organizations would be needed to carry out evaluation and selection. Corps members would be selected for specific terms, perhaps five years, with renewal requiring re-competition.

2. **Size**. Ultimately, we would favor a STEM Master Teachers Corps that recognizes and rewards the top 20 percent of STEM teachers in the Nation. This proportion might act as a sufficient carrot to attract and retain the best teachers. In the initial stages, we would favor a program that is smaller – but still much larger than the tiny programs that currently exist. An appropriate initial proportion might be the top 5 percent of high school and middle school STEM teachers, together with some extraordinary elementary school STEM teachers, distributed geographically across the Nation. The total might be about 22,000 teachers. This would correspond to roughly 50 per Congressional district, which could be an appropriate way to distribute membership.
3. **Recognition and Respect.** Members of the STEM Master Teachers Corps would be recognized and respected as a professional elite. The importance of such recognition cannot be overstated in terms of attracting and retaining excellent teachers. The recognition also would drive demand for excellent teachers by schools, school districts, and parents. Moreover, the existence of a Corps would create a national network of teachers who could serve as mentors, leaders in their schools, and liaisons to public officials. Social networking and interactive multimedia technology could connect these teachers and their classrooms and help them learn from each other and mobilize groups toward common aims. Excellent teachers would be connected with their peers and could share best practices and materials. It would expand teachers' opportunities for professional growth while keeping them in teaching by giving them leadership status within their schools and among other teachers across the Nation. They would also have a voice for their profession in education policy and could be effective advocates for STEM in their schools, school districts, and communities. The Corps has the potential to have a powerful multiplier effect on STEM education.
4. **Salary supplement and other resources**. In addition to recognition, members of the Corps should receive a significant salary supplement. The amount of salary supplement needed to make a significant difference in attracting and recruiting excellent STEM teacher would need to be determined based on economic analysis. Based on the market considerations above, we suspect that a supplement in the range of $15,000 per year would be necessary and appropriate. If so, the program might cost in the range of $325 million per year. It is possible that some matching support could be raised from the private sector. It may be appropriate for the salary supplement to be greater for teachers in high-needs schools to provide them with an incentive to stay in these schools. Finally, in addition to salary supplementation, these teachers also should receive some discretionary resources for use in their classrooms.

   Recognizing a large number of teachers across the country will establish a professional elite that defines aspirations for STEM teaching. It will send a signal to all STEM teachers that their profession is respected and will encourage them to strive to meet a nationally recognized high bar of excellence. It will promote their interaction as part of a national network and bring their wisdom to bear on improving local and state policies that crucially affect STEM education.

If the program were to encompass 50 teachers per Congressional district, the annual cost would be approximately $750,000 per district, yielding the total cost of approximately $325 million per year. Because the five-year terms would be staggered, the programs costs would ramp up linearly over five years. In addition, effectively implementing this program across the Nation will require full-time professional staff in each Congressional district who can network teachers to each other and help provide a voice for those teachers on education policy issues.

(Currently, the Presidential Awards for Excellence in Mathematics and Science Teaching, administered by the NSF, disburses $1 million per year to 100 teachers. Substantially more funding will thus be required to recognize and reward a meaningful fraction of teachers in a STEM Master Teachers Corps.)

The six-month study on implementation should include among its range of considerations: the degree to which regional organizations or consortia might administer the STEM Master Teachers Corps, the balance in the selection criteria of taking into account teacher content knowledge and pedagogical skills as well as student outcomes, and the potential to build on existing Federal and state recognition programs for teachers.

Recommendation 5-3. Create a STEM Master Teachers Corps

As a powerful way to promote excellence among the Nation's STEM teachers, the Federal Government should support the creation of a national STEM Master Teachers Corps to recognize, reward, and engage the best STEM teachers in the Nation and elevate the status of their profession.

The STEM Master Teachers Corps should initially consist of the top 5 percent of STEM teachers distributed throughout the nation, primarily at the middle and high school levels, and should be selected based on their demonstrated ability to prepare and inspire students.

Corps members should receive significant salary supplements (in the range of 20-25%) of their salary, as well as some funds to support activities in their schools and districts. They also should be provided with opportunities to interact with each other and to have a voice in education policy at the district, state, and national levels.

Because various issues must be addressed in establishing a STEM Master Teachers Corps (such as criteria and process for selection, program structure and administration, and funding sources), the Department of Education and National Science Foundation, with coordination from the Domestic Policy Council, undertake a six-month study of how best to implement the corps.

The Federal Government also should seek to engage companies and philanthropists to help support the STEM Master Teachers Corps, because STEM master teachers represent national treasures worthy of respect.

We recommend that the Federal Government undertake a rapid six-month study to address the issues in implementing a STEM Master Teachers Corps – including the selection process and criteria for the teachers and the organization and administrative structure

VI. EDUCATIONAL TECHNOLOGY

We are now...at an inflection point for a much bolder transformation of education poweredby technology.

–National Educational Technology Plan, 2010, Department of Education[149]

Chapter Summary

Technology has the potential to transform K-12 education, just as it has many other sectors of our economy and our society. It can enable real-time and meaningful data-gathering that allow learning and innovation in the education system. It can power innovative learning tools that prepare and inspire students. And all this can be done while empowering, rather than replacing, teachers and traditional teaching methods. Thus far, however, the influence of technology on schools has been limited. A new mission-driven approach to the development of technology platforms, "deeply digital" and comprehensive instructional materials, and powerful new assessments could dramatically improve the STEM preparation and inspiration of all students, including those at risk of losing interest in STEM subjects. The Federal Government should establish a mission-driven R&D entity to drive widespread use and scale-up of innovative and effective technologies in the K-12 education system. It should nurture a vibrant ecosystem of for-profit and not-for-profit enterprises that will bring innovation to K-12 education and prepare and inspire students in new ways.

Introduction

One of the most powerful ways to improve K-12 education is to instill a culture of innovation throughout the system. A feature of most great organizations, a culture of innovation encourages creative experiments, collects data, measures results, recognizes improvements, and rewards success. Data-driven feedback fuels steady improvement by allowing people and institutions to test new ideas and learn from their successes and their errors.

The Obama administration has demonstrated its commitment to instilling a culture of innovation in the Nation's K-12 education system, through its focus on the collection and use of data, and through such programs as the Race to the Top and the Investing in Innovation Fund, which provide funds for scaling up proven programs. These programs and others have begun to help education systems embrace the possibility and importance of innovation.

PCAST believes that one of the most powerful tools to propel innovation in education is computation and information technology. As described in this chapter, we believe that computation and information technology is today poised to play a transformative role in education, and that wise investment in technology-based initiatives can yield enormous benefits. These views are consistent with those expressed in the Department of Education's recently released draft National Education Technology Plan, which contains observations and recommendations with which we concur.

Over the past two decades, computation and information technology has been a powerful driving force for innovation in many fields and has transformed our world. By providing increasingly ubiquitous access to connectivity, knowledge, and analytical power, technology has unleashed a revolution in the way we live, work, and play. Vast libraries of the world's knowledge are now available from our cell phones. New communities have sprung up through social networking, wikis, and real-time blogging. Individuals have greater opportunities to fulfill their talents and express their opinions. Technology-based innovations have created economic efficiencies that allow high-quality products to reach larger markets with lower costs, connect buyers and sellers in online auctions, support "just-in-time" inventories, and enable flexible design and production. Entire new industries have been created.

Technology supports innovation in three fundamental ways:

- **Continuous evaluation and improvement based on data**. Technology-based initiatives facilitate collection and evaluation of data, providing rapid feedback to test and improve ideas. Google constantly refines its page-rank algorithms by studying which links users follow, and Netflix similarly refines its movie suggestions; these companies learn from millions of data points each day. Book publishers can test the appeal of titles and covers, and marketers the appeal of their ads. In short, the ability to gather and interpret data promotes a culture of learning.
- **Rapid and inexpensive dissemination of successful solutions**. Technology-based solutions have the advantage that successful innovations can be rapidly and inexpensively disseminated to students and schools, as long as the necessary technology is available and they have reliable connections to technology services.
- **Mass customization**. Technology can support customization, such as instructional materials that include different approaches suited to different levels and learning styles and problem sets that adapt to student responses.

Despite its transformative role across the global economy, technology has *not* played a major role in K-12 education to date. As we discuss below, there are understandable reasons why this has been the case. Nonetheless, this represents a tremendous missed opportunity.

PCAST believes that technology has now advanced to the point that it can and should play a leading role in driving educational innovation. Moreover, we believe that STEM education should be the leading wedge for such change. In this chapter, we therefore recommend that the Federal Government launch a major new initiative in educational R&D tied to information technologies. Our recommendation is rooted in several observations.

- First, educational technology has been advancing rapidly in recent years and is likely to create major strides in the near future. From a scientific and technological standpoint, there is unprecedented ferment in the field.
- Second, there will be a growing need for new instructional materials, new professional development materials, and new kinds of assessments that are aligned with higher standards and provide much richer learning experiences and more vibrant sources of information. To that end, technology has the ability to shorten innovation cycles, expand the market of suppliers, and assess learning in deeper ways. It also has the capability to respond to the diverse learning styles of students.[150]

- Third, the "collection and use of data" is one of the Department of Education's four assurances. Technology is a powerful tool in support of this goal, because it provides efficient ways to gather, integrate, and analyze rich and diverse data streams to evaluate and improve programs.
- Fourth, technology is becoming increasingly affordable, accessible, and versatile. This trend will continue over the next decades, and will encompass personal and mobile devices.
- Fifth, today's students are increasingly digital natives. They are used to technology and have come to expect high-quality uses of technology.
- Sixth, instructional materials for students and teachers developed in the United States could have uses throughout the world, bringing together students and educators in partnerships and activities that will benefit all participants and contributing to the Nation's outreach to the next generation of students throughout the world.
- Finally, while the provision of education is the province of the states, only the Federal Government has the ability to fund the basic R&D necessary to develop truly transforming platforms and instructional materials for education. State and local educational authorities lack the necessary resources or scale.

The National Educational Technology Plan (NETP)[151] and the National Science Foundation's report on Cyberlearning[152] also describe the unparalleled opportunity for technology to play a transformative role in K-12 education.

Before launching into a detailed discussion of our recommended strategy, we underscore three crucial points:

- **Technology cannot replace the need for great teachers**. The goal of technology is not to replace teachers, but to support them. Properly used, technology can extend the reach of teachers by giving them access to the best instructional and professional development materials, to tools that can create customized learning environments and assessments for students, and to data that capture rich information about individual performance.
- **Ensuring that schools and students have adequate technology infrastructure will require considerable attention and resources**. Given the direction in which technology is moving, we believe that R&D should be aimed at a world in which there is universal access to computing devices and to connectivity. Nonetheless, we must keep in mind that we are not yet in this world. Many schools lack widespread access to computers, up-to-date software, adequate broadband connectivity, and adequate maintenance budgets. While the situation has improved and will continue to improve as costs fall, this will require major attention. The Federal Communications Commission's recent broadband plan[153] and the National Educational Technology Plan address these issues, and changes need to be made through appropriate legislation.
- **Technology-based solutions will need to be rigorously and continually evaluated to assess whether they are increasing student preparation and inspiration**. While we believe that technology is a critical component in achieving the Nation's education goals, technology-based approaches will vary in their effectiveness and

must be subjected to careful evaluation. Fortunately, technology makes it possible to collect rich streams of data. Technology-based projects should be designed from the beginning with evaluation in mind.

Early Efforts in Technology in K-12 Education

Efforts over the past two decades to employ technology in K-12 education have had less impact than expected. There are various reasons why this has been the case, and it is important to understand them in charting a path forward:

- Some efforts to introduce technology in schools have been misguided in that they aimed to replace rather than support teachers. Moreover, many efforts have not taken into account the degree of preparation and guidance needed to help teachers use technology in the classroom.
- Some technology-based learning products have had mediocre quality, uninspiring content, and lackluster production values. Educational applications require greater sophistication and higher quality than office productivity tools. One problem has been that adequate tools and infrastructure (such as adaptive software and adequate bandwidth) were not available until recently. Another problem has been that market forces in education have not been strong enough to call forth the best technology development. Today's remarkably sophisticated video games stand as a marked contrast of what can be achieved when there is adequate market demand.
- Public schools have historically had poor access to technology infrastructure. Even today, computers are often confined to a "computer lab" and teachers lack their own computer. While most schools today report that they have Internet access, the connections often have narrow bandwidth shared by too many computers. A challenge that persists is the lack of necessary support staff, resources, and expertise to maintain, update, and repair technological infrastructure in schools over time.[154]
- While many instructional resources are available on the Internet, they tend to be unorganized and fragmentary. Teachers lack the time to sift through endless amounts of content to find and integrate high-quality, proven materials into lessons. Efforts such as the National Science Foundation's National Science Digital Library have made some progress on this problem, but they fall far short of what is needed. Most importantly, teachers need coherent whole-course materials, together with teacher support, that are based on solid research and take full advantage of information technologies. Currently, there are few well-tested, coherent whole-course units that teachers can use.
- Funding for technology-enabled educational innovation has been limited. Moreover, technology- based programs are at a competitive disadvantage in competing for grant funds because they typically require greater investment. Just 12 percent of the funding distributed by NSF's Directorate for Education and Human Resources goes to developing new technology, applying technology in innovative curricula, or using technology for student assessment.[155] Also, the average grant for applying educational technology was only about $1 million, which must be used for

curriculum development, testing, research, external evaluation, and dissemination, leaving little funding for the development of the underlying technology.

- Innovation has been limited by a lack of common technology protocols and platforms. Instructional materials are often designed to work with a particular company's unique learning system, and these systems are typically incompatible with other systems. This restricts access for educators who want to be able to select among the best materials and discourages innovators, who are attracted by access to wide markets.

Recent Progress

Box 6-1. Technology-Based Approaches for Education

Technology-based approaches are beginning to have an impact in education. Most are still at early stages, but they point to a transformative potential.

Wikipedia and the power of communities. Wikipedia and similar projects have shown how worldwide communities can author, edit, and update vast libraries of content and make knowledge and tools globally accessible. Similarly, social networks can connect groups of professionals and students devoted to learning from each other, creating opportunities for teachers and students to participate in interactive learning environments that extend far beyond the classroom. These forums create the challenge of vetting vast amounts of information for high-quality, useful resources, but various approaches (such as curation by experts and general users) are increasing the capability of these tools to disseminate ideas, innovations, and best practices.

Open courseware. Universities such as the Massachusetts Institute of Technology and the University of California, Berkeley, have provided free online access to higher education courses in STEM and other subjects, making lectures and associated materials available to learners of any age anywhere in the world. These materials can have a particularly powerful impact on students in developing countries who access them. They can empower both students and teachers to learn beyond the curriculum and beyond the school day. Although videos of lectures are not the most dynamic way to engage students in virtual learning, they represent a promising first step toward sharing educational content. Working together, college and K-12 educators could generate short lectures and demonstrations with higher production values that are appropriate for various levels.

Electronic textbooks. Electronic textbooks have many potential advantages over hard-copy versions. Even if they are simply electronic images of hard-copy text, e-texts can be cheaper to purchase, lighter to carry, easier to update, more readily adapted for English-language learners, and more accessible to students with disabilities. They can enable innovative companies and foundations to contribute to making high-quality educational materials. Driven by budgetary constraints, California, for example, has recently launched an initiative to create open-source, free e-texts for high school mathematics and science, and a nonprofit has created several textbooks that meet the state's requirements for public school classrooms. At their best, e-texts can provide a richer and more engaging experience through interactive components, videos, simulations, hyperlinks to additional resources, graphics elements, assessments that gauge student interest and comprehension, and much more.

Tutoring systems. New types of computer-based tutors can lead students through STEM materials that include ongoing and adaptive assessments. In this way, tutors can diagnose student difficulties and suggest activities that address those difficulties. This approach appears to work best for teaching structured content such as mathematics through algebra and introductory computer programming. For example, the Cognitive Tutor, which grew out of Artificial Intelligence research at Carnegie Mellon University, has yielded substantial improvements in student achievement.[157]

Online courses for students. Online courses are being developed by a variety of organizations, including a growing collection of "virtual schools" that allow students to take individual courses or their entire school program through online encounters. Some statewide entities are already experimenting with virtual K-12 schools. The Florida Virtual School (http://flvs.net), for example, offers a range of multimedia content and online courses taught by certified teachers for K-12 students. The Colorado Virtual Academy (http://www.k12.com/cova), the Pennsylvania Virtual Charter School (http://www.pavcsk12.org), and Virtual Virginia (http://www.virtualvirginia.org) similarly provide students with online learning experiences. More than one million students are now served by online classes, covering core and remedial courses.[158] STEM courses with laboratories can be difficult to put online, but simulations can reproduce aspects of the laboratory experience. Online courses need to be evaluated to see what is most effective.

Online professional development. Online programs have begun to emerge that can help teachers engage in in- service and off-site professional development. A pioneer in this area, the National Teacher Enhancement Network, has been offering online, credit-bearing professional development courses in science since 1993. PBS TeacherLine (http://www.pbs.org/teacherline/) and Teachscape (http://teachscape.com) are other examples of online resources through which teachers can improve their content and pedagogical knowledge. Several universities offer programs ranging from individual short courses to full degrees online.[159] Importantly, not all online professional development programs are of high quality. It will be important to develop ways to assess programs based on outcomes and certify those that are effective.

Assessments. Information technologies offer incredibly powerful ways of assessing student's thinking while they are learning so that teachers can create more effective learning experiences. For example, the National Assessment of Educational Progress has been exploring the use of technology-based simulations, such as determining the relationship between payload masses and the altitude of a helium balloon, to measure students' computer skills, inquiry skills, and synthesis skills. [160] Technologies also can collect, analyze, and disseminate data about student learning and make this information available to educators, schools, districts, states, and the Nation to support continuous improvement in STEM education.

Integrated learning sites. Wolfram |Alpha (http://www.wolframalpha.com) is a free "knowledge engine" that generates immediate answers to questions by mining its internal database in real time. Wolfram |Alpha differs from conventional search engines by supplying an explicit response to each query rather than a list of sites through which a user must sift to find the information of interest. It also hosts tools of use in the classroom setting, such as online lesson plans and demonstrations, and offers apps for common technologies like the iPhone. Educators, STEM professionals, and students have developed modules for the site that are made publicly available after review for quality control.

Cloud computing. One of the challenges for schools and school systems is hosting and maintaining software and systems. Some businesses and universities have begun to use cloud computing services, in which data and programs reside in off-site repositories, to reduce the cost of maintaining and repairing expensive on-site equipment. School systems may benefit from such solutions.

Games and simulations. Video games and detailed simulations have achieved levels of sophistication that make them into compelling tools for learning inside and outside of classrooms. They require and foster many 21st-century skills such as systems thinking, problem solving, information tracking and resourcing, collaborative information sharing, leadership, teamwork, and communication.[161] For example, Whyville_(http://www.whyville.net) is a virtual world populated by 5 million children, with more girls than boys, that offers science- and math-based activities targeted at 8- to 14-year-olds.[162] The Department of Defense has also made considerable use of simulations and games for training.[163]

While technology has had only a modest impact on the K-12 classroom to date, the current level of activity and creativity in the world of educational and learning technology illustrates its tremendous potential. The examples range across the entire educational spectrum, with some of the most creative occurring at the college and university level. We cite some examples that illuminate these trends, including ones that apply across all fields of education (see Box 6-1) and ones that are specific to STEM subjects (see Box 6-2). While none of the examples fully addresses the needs of K-12 education, together they indicate how far reaching and disruptive educational technologies could be.

Box 6-2. Technology-Based Approaches for STEM Subjects

Technology-based materials can be particularly powerful for STEM subjects, because they can assist in visualization, experimentation, and data collection.[164] Scientific methodology and culture are rapidly changing to a digital, collaborative mode. The approaches that have emerged in STEM fields should be shared with today's students and featured via innovative technologies and STEM-focused schools.

Math visualization tools. Visual representations of abstract mathematical concepts can produce learning gains that are difficult to achieve in other ways. Graphing software using both calculators and computers is already an accepted part of algebra instruction. These visualizations barely scratch the surface of what is possible. For example, the SimCalc project [165] makes the "mathematics of change" come alive with computer visualizations; it could easily be incorporated into mathematics starting in elementary grades to provide a foundation for calculus. Similarly, interactive software can be used to construct and explore principles of geometry, allowing students to discover and prove theorems of geometry themselves instead of reproducing textbook proofs. GeoGebra, a free open-source mathematics visualization tool for high school subjects, and Scratch are two other examples of promising programs.

Probeware for laboratory experiments. Probeware – systems that allow sensors to feed data directly into a computer for immediate analysis and display – allows experiments to be performed more quickly, frequently, and flexibly. By providing instant feedback, it helps students understand the importance of experimental design and the significance of data. Developed at TERC in the 1980s, probeware now includes inexpensive sensors capable of measuring everything from acceleration to energy flows. One vendor, for example, has developed an application that wirelessly links probes that measure the tensile strength, toughness, and ductility of materials to the iPad and iPhone, where students can measure and interpret their results. [166] The Internet System for Networked Sensor Experimentation (iSENSE) enables users to contribute data collected using classroom probes and other sensors, view and analyze data from other contributors, and combine data from multiple sources to examine regional, national, and global phenomena.[167]

Software for virtual experiments. When real experiments are not feasible for reasons of cost, safety, time, or scale, interactive computational models with graphical outputs can simulate the experience. For instance, the Molecular Workbench simulates the motion of atoms and molecules under good approximations of atomic-scale forces and interactions.[168] It can be used to explore a wide range of phenomena that depend on atoms and molecules, such as phase change, thermal expansion, diffusion, osmosis, activation energy, black body radiation, and protein folding. Depending on how the activity is designed, the program is suited for students from elementary grades though college.

Programming as a tool for teaching computational thinking. One of the earliest educational applications of computers involved engaging students in programming using simplified languages. The language Logo was invented in 1967 and used to control a real or virtual robot, called a "turtle," through simple instructions. Logo spawned multiple generations of offspring programming environments, [169] such as Lego's Mindstorms robotics product used in robotics competitions and NetLogo, an agent-based version of Logo, in which many parties can interact.[170] Another example is Alice (http://alice.org), an innovative object-oriented 3D programming environment with which students can create an animation, play an interactive game, or prepare a video to share on the web.

Student collaboration. Networked computers allow students to interact with STEM professionals as well as other students worldwide to carry out class projects. For instance, iEARN[171] is an international network of more than 30,000 schools in over 130 countries with dozens of curriculum-linked STEM projects for all grade levels, most contributed by teachers. The BioKIDS/Deep Think project at the University of Michigan is using inquiry-based examinations of organisms and ecosystems to improve learning in high-poverty elementary and middle schools in Detroit.[172]

Missing Pieces

Despite the tremendous activity and advances in educational technology, critical pieces missing from the picture are impeding the realization of its full benefit for K-12 education:

1. **Coherent integration**. Most of the applications described above and many like them have to be integrated by teachers into their instruction. Some teachers who are early adopters do this routinely, selecting materials they feel fit their students' needs and their own instructional goals and preferences. But most teachers lack the time, confidence, content knowledge, and inclination to do so.
   There is currently intense interest – and some preliminary work—combining the best of all the technologies and concepts sketched above into larger, coherent curriculum units and complete courses. These "deeply digital" materials would use simulations, probes, multimedia, and other digital resources as needed. An example of this work is the WISE project at the Technology Enhanced Learning of Science Center,[173] which has a platform that has been used to create hundreds of learning activities and a public research component.
   The ideal instructional package might include a mix of modules and types of applications that could provide alternative paths for different learners. It would have built-in assessments with ongoing feedback to the teacher and, optionally, an AI tutor. It would support teachers in doing what they do best such as leading discussions and evaluating student work, while leaving time for students to explore individually or in small groups. The software would include supports for students with special needs, vocabulary and pronunciation aids, and intelligent help. These materials might support modest increases in class size while increasing student learning.
2. **Technology platforms and standards for interoperability.** A key to realizing the promise of technology for education is the availability of open-source technology platforms and standards to allow interoperability. By separating the creation and the delivery of content, technology standards and platforms have unlocked the creative potential of the Nation in other domains; examples include the World Wide Web and the iPhone.

In education, content creators face the challenge of developing materials that interface with other educational technology systems. Schools face the challenge that many educational technology systems are based on proprietary formats that cannot exchange or integrate data. The availability of common technology platforms would lower the barriers to entry by innovators and the costs to schools. They would make it easier to author and deliver courseware on a wide range of devices. They also would enable schools to draw from a wide range of sources in constructing learning programs that meet their needs.

Such courseware needs the ability to perform a wide range of functions. For example, it needs to deliver and sequence pages; handle student interactions with elements on these pages, track student progress; support classroom interactions between peers, the teacher, and remote classes; and collect data on student performance for the teacher and researchers while controlling access to the data to protect privacy.

Open-source versions of platforms and materials would greatly reduce the cost and increase the speed of developing such courseware. Publishers would be able to copyright and market materials made with these technologies, creating a vibrant market in deeply digital materials. With open-source platforms and standards, even small players could create new modules. Such an environment would allow developers to mix-and-match components to support continuous improvement.

Early efforts to develop common technical standards include the Shared Content Object Reference Model (SCORM) developed through the Advanced Distributed Learning (ADL) Initiative of the Department of Defense and the more recent Common Cartridge framework developed by the IMS Global Learning Consortium. However, much more work is needed to develop the required common infrastructure.

3. **Connectivity and Computer Infrastructure.** Although almost all schools now have some form of connection to the Internet, the levels of connectivity are often not sufficient to meet current demands and could fall further behind demand in the future.[174] Connection speeds are not high enough in many schools, and new technology-enabled materials will require greater bandwidth and more sophisticated tools. Many schools will require significant upgrades to meet future demands, and schools in rural or high-poverty areas are at risk of falling farther behind more advantaged schools.
4. **Market demand to create a vibrant ecosystem of technology-based innovation.** Despite the great potential for educational technology, an important challenge is that the current market for educational technology tools and services is fragmented. This fragmentation discourages private capital and innovative companies from investing in the field. It also inhibits digital materials and devices from being produced and distributed at a scale that would drive their prices down for schools and school systems. The decision makers who purchase technological tools and materials in the K-12 public school system in the United States span 50 state education agencies, 14,000 school districts, 98,000 schools, and more than 2 million teachers. They must follow regulations that constrain how they spend public funds, and they often lack the resources to maintain and update their technology infrastructure. The market today is not equipped to foster a thriving R&D enterprise for education technology.

The broad adoption of the Common Core academic standards is a potential watershed for improving the dynamics of demand in K-12 education. A product created to serve one state could now, with minimal additional investment, serve multiple states. This lowers the cost of entry for new educational technologies and makes the market more attractive for investors.

In addition, considerable progress could be made through greater collaboration among states and districts in the evaluation and purchase of products. The Achieve Algebra II consortium and the Race to the Top assessment program have established precedents for multistate collaboration in assessment. Similar consortia could aggregate evaluation and demand in ways that attract substantial private capital. In addition, modular design of instructional materials could lower barriers to entry for small developers.

A Vision for Technology-Driven Innovation in K-12 Education

Based on ongoing advances in information technology, PCAST strongly believes that technology initiatives should be a central agent of change in K-12 STEM education. These initiatives should draw on the following vision:

1. **Deeply digital, whole-course instructional materials with several alternative versions for all major STEM courses.** These materials should make optimal use of text, images, videos, interactive simulations, games, collaboration tools, embedded assessments, and adaptive problem engines. The materials should take advantage of and provide access to rich data sets and computational techniques available in STEM research communities. They should be designed in a modular fashion so that they can be readily expanded, changed, improved, adapted, and updated to fit different needs and settings. Because the majority of STEM education at the middle school and high school level falls into a few dozen courses, it should be feasible to create at least several alternative versions for each.
2. **Modular components for use in instructional materials**. STEM practitioners in universities and industry should be able to contribute modules that can be combined into whole courses so that parts of the course can, on the basis of ongoing research, be changed and continually improved. These modules could consist of images, short videos, simulations, and data sets. Many STEM practitioners who are not in a position to create whole-course materials would be eager and able to contribute superb components, often in an open-source manner through Wikipedia-like mechanisms. These modules could be invaluable tools for developers of instructional materials. Incentives for the creation of such modules could include grants, prizes, and recognition.
3. **Testing systems and test materials.** Assessments should enable data mining and collection of student performance data on an ongoing basis, not just at the end of a school year. Such systems also should collect data on student and teacher experiences with curricular materials with which to improve materials on a continual basis.

4. **Personalized tutoring that extends beyond the classroom.** Online tutors who can customize the pace and delivery of material can help students who are struggling by providing individualized attention and instruction that is accessible 24/7. Accelerated students also should have access to enriching learning experiences and challenges so that their knowledge and achievement grow over time.
5. **Automated systems and software to aid teachers.** Educational technology can save teachers time by helping them grade, get student feedback, gauge how particular topics should be taught, and create lesson plans. Such systems should be easy to use, integrated with digital curricular materials, and accompanied by training modules for teachers. They should facilitate teachers' creation of online learning communities for their students, as well as networks with other teachers and with STEM professionals.
6. **Technical platforms, standards, and exchange languages for digital educational materials, data, and services.** As discussed above, common infrastructure will lower the barriers to innovation, promote competition, lower costs to schools, and facilitate data collection and analysis.

Need for an Advanced Research Projects Agency

Ultimately, the impact of information technology on K-12 education will be realized through the efforts of developers of instructional materials, computer scientists, publishers, technology providers, teachers, schools, and local and state education authorities. But the Federal Government has a unique role to play: Only the Federal Government is in a position to support the underlying R&D needed to drive this transformation.

Specifically, the Federal Government can and should support:

1. The development of interoperable, open-source technologies upon which instructional materials can be constructed for all subjects.
2. Research on the educational impact of STEM materials based on these technologies.
3. Development of initial versions of coherent, research-based, course-length materials for all STEM areas (which would give rise to specific products marketed by large and small educational entrepreneurs).
4. Research on new kinds of assessments and new ways to develop assessments.
5. Mechanisms to engage public and private sector researchers to address grand challenges for education R&D, such as those described in the National Educational Technology Plan.

The participation of for-profit companies, nonprofit organizations, and government agencies in the development of interoperable, open-source technologies will propel continuing technical innovations. Competition to develop and market whole-course instructional materials based on these innovations will ensure the wide dissemination and continued support of the best materials at reasonable costs to schools.

The Federal Government has sought in the past to propel educational technology in general and with respect to STEM subjects, but these have fallen short of what is needed.

- The National Science Foundation launched the National Science Digital Library (NSDL) in 2000, with the aim of collating various educational materials related to STEM in a web-based portal for teachers. However, it does not sufficiently address the issues of interoperability or teachers' need for coherent, easy-to-find, whole-course instructional materials. It is online, but its materials are not necessarily technology based. It has not marshaled investment from the private and public sectors in mission-driven R&D to develop innovative technology-based platforms, assessments, and materials for K-12 education and has only a limited impact on education.
- The Higher Education Act of 2008 authorized the creation of the National Center for Research in Advanced Information and Digital Technologies, also known as Digital Promise. Digital Promise was structured as an independent nonprofit foundation housed within the Department of Education to undertake precompetitive education technology research and development. With sufficient resources, it could play a useful role in fostering interagency collaboration on education R&D and private-public research initiatives. But Digital Promise cannot achieve the goals discussed in this chapter. Its design and governance structure are not suited for a mission-driven focus that will support project-oriented educational R&D to produce technology platforms, whole-course curricula, assessments, and learning systems that can be brought to scale in the K-12 school system.

To accomplish the goals laid out in this report, the Federal Government needs a framework to support mission-driven R&D in K-12 STEM education. We are convinced that this work cannot be done through typical grants from the Department of Education or the NSF's Education and Human Resources Directorate. The culture of grant review and supervision is not well suited to the needs. Instead, an entity is required that has the knowledge and culture to manage technology investments in a mission-driven manner, including development of technology platforms and large integrated projects. The entity must be driven by a mission to improve educational technology and to spread innovations into schools and classrooms.

We believe that the best solution is to create an entity that we will refer to as the Advanced Research Projects Agency for Education (ARPA-ED), with would work collaboratively with both the Department of Education and the NSF. The entity could reside in one of these agencies or as an independent entity governed by both. Regardless, it must have the culture and the leadership to spearhead mission-driven educational innovation and to achieve widespread results.

RECOMMENDATION 6-1. CREATE AN ADVANCED RESEARCH PROJECTS AGENCY FOR EDUCATION (ARPA-ED)

The Federal Government should create an entity, which we will refer to as an Advanced Research Projects Agency for Education (ARPA-ED), with the primary goal of propelling and supporting (i) the development of innovative technologies for learning, teaching and assessment across all subjects and ages, and (ii) the development of effective 'deeply digital' whole-course instructional materials for

STEM education that prepare and inspire the next generation of American students. (See Box 6-3 for a fuller description of ARPA-ED activities.)

ARPA-ED should have a mission-driven focus to engage in directed development projects with the potential to produce revolutionary improvements. The entity should have the ambition, accountability, authorities, and flexibility that have historically characterized the Defense Advanced Research Projects Agency (DARPA). It should aim to ensure that high-quality digital technologies and instructional materials are put into widespread use in K-12 classrooms.

ARPA-ED should be tightly coupled to both the National Science Foundation and the Department of Education, and, especially in its early years, it should also work closely with DARPA. The precise organization structure should be worked out by the administration, but it should ensure that ARPA-ED has strong leadership and a strong culture that is distinct from existing grant-making programs at NSF and ED. ARPA-ED should be initially funded at a level of $100 million, and will eventually require a budget in excess of $200 million per year. ARPA-ED should also work closely with visionary corporations, universities, and foundations to accomplish the goals, including in the development of instructional materials.

We recommend that the Department of Education and National Science Foundation, with the coordination from the Domestic Policy Council, undertake a rapid six-month study of how best to implement ARPA-ED, including examining how the agencies could begin the work using existing authorities and funds. We also recommend that ED and NSF convene a series of workshops with stakeholders to identify the most important initial projects.

BOX 6-3. ARPA-ED GOALS AND ACTIVITIES

ARPA-ED should nurture a vibrant ecosystem of for-profit and not-for-profit organizations, serving individual learners and state and local education authorities, by providing support for such activities as:

- **Development of common technology platforms, to facilitate development and integration of content from diverse sources and data exchange among platforms.**
- **Development of technology-based learning platforms that can readily combine text, short videos, lectures, demonstrations, and interactive simulations into coherent content in a modular manner that can be readily customized or extended; that employ adaptive content and tutoring systems to tailor learning to each student; and that incorporate systems based on student and teacher feedback and on student performance to allow continuous improvement in the pace and effectiveness of learning.**
- **Development of "deeply digital" whole-course materials for K-12 STEM education, including instructional materials; homework problems; research problems; formative and cumulative assessments; and professional development materials. The goal should be to create multiple alternative whole-course versions for all STEM courses, including both in-class courses for use in schools**

and self-paced, self-directed courses that allow students to achieve or surpass STEM curricula in non-traditional ways.

- **Development of modular instructional materials by the STEM community, such as simulations, videos, and research problems, encouraged and incentivized through novel mechanisms such as prizes.**
- **Development of innovative assessments that could be created more efficiently and could test understanding more deeply.**
- **Rapid prototyping and evaluation of systems and materials in classrooms.**
- **Data-mining initiatives, in partnership with states, districts, and consortia of school, to use research on anonymized longitudinal data to improve educational practice.**
- **Pre-competitive consortia to stimulate research and development of pre-competitive technologies of broad importance.**
- **Innovative procurement consortia of state and local education authorities to aggregate demand for new learning technologies and approaches.**
- **Innovative ways to ensure broad dissemination and effective uses of tools and content, including through availability of open-source reference implementations and development of open-access materials.**

Jumpstarting ARPA-ED. While the formal creation of ARPA-ED may require time, we believe that the Department of Education and the National Science Foundation can begin this work through an interagency collaboration using existing authorities and funds. Some near-term possibilities include:

- Using a portion of the Investing in Innovation (i-3) funds at the Department of Education to support ARPA-Ed activities.
- Aligning the goals of the NSF's Cyberlearning and Transforming Education Initiative with the goals of ARPA-Ed.
- Utilizing the National Center for Research in Advanced Information and Digital Technologies (Digital Promise) to launch a public-private partnership focused on high priority opportunities.

There are advantages and disadvantages to each of these approaches, and we therefore recommend that the agencies develop a plan to ensure the long-term success of ARPA-Ed while undertaking the intensive 6-month implementation study recommended above. We also encourage the Department of Education and the National Science Foundation to consider how to engage the other science mission agencies, in particular the National Institutes of Health, which is developing technology-based learning tools, in the work of ARPA-Ed.

It is also important to identify key initial projects, including specific technology platforms and particular courses for the development of instructional materials. For this purpose, ED and NSF should convene a workshop including state and local education authorities; education publishers; innovative companies that create platforms, modules, games, virtual communities, and instructional materials; experts in learning science; technology industry leaders and innovators; teachers; school administrators; parents; and students.

In parallel with this heightened R&D activity, it is important that the Congress support the administration's initiatives to ensure that the Nation's schools have adequate infrastructure for technology, as laid out in the administration's National Broadband Plan and the National Educational Technology Plan. Over time, the plummeting costs of technology and networking and the availability of comprehensive learning materials will enable most schools to provide adequate technology. Eventually, the total costs of technology-based approaches will become comparable to textbooks (they are currently estimated to be two- to-three-fold higher), at which point schools will move naturally toward such solutions. At all stages, however, it will be important to ensure that low-income schools are not unfairly disadvantaged. In addition, it is critical that programs provide long-term information technology support for schools to maintain, repair, and update their technological tools and that teachers be provided with technology training.

VII. Students

There exists a passion for comprehension, just as there is a passion for music. That passion is rather common in children, but gets lost in most people.

– Albert Einstein

Chapter Summary

STEM education in K-12 relies on our ability to motivate and inspire students. This involves creating exciting opportunities for students to have individual or team-oriented experiences with the ideas, discoveries, and emerging knowledge in STEM fields. Two key avenues can provide students with experiences that allow them to explore and challenge themselves with STEM. The first is out-of-class and extended day activities that include contests, laboratory experiments, field trips, and more. The second is advanced courses that press students to set ambitious goals and achieve at higher levels. All students should have opportunities to have these two kinds of experiences. In particular, girls and members of minority groups underrepresented in science and engineering can find inspiration in these activities and mentors and role models that encourage them to study STEM subjects and enter STEM professions. The Federal Government should launch a coordinated initiative for out-of-class and extended day programs in STEM. Existing after-school programs should be reoriented toward STEM experiences, and support for contests and programs in STEM should be increased. In addition, the Federal Government should take steps to ensure that many more students take and succeed in advanced classes in STEM subjects that go beyond the standard curriculum.

Introduction

Every child is a scientist, it is often noted.[175] Children are naturally curious and creative. When students realize that they can discover new things about the world and construct explanations about how the world works, discovery can become a deeply personal and lifelong passion. STEM education should build on the proclivities of young people to think carefully about profound issues, solve problems that pose risk to human societies, create and fix mechanical objects, observe or understand phenomena that no one has observed or understood before, probe the behaviors of people, or any of a wide variety of other activities.

These personal experiences are an important source of inspiration and learning for students. They demonstrate the power and applicability of the material taught in mathematics and science classes. They show how continued study in these fields can lead to meaningful and productive careers. They give students a reason for studying STEM subjects and high goals toward which they can strive. They help incite a passion for STEM that can fuel a career and last a lifetime.

Research shows that boys and girls who show interest in STEM in eighth grade can be three times more likely to later pursue degrees in STEM fields. This underscores the importance of giving children exciting opportunities in STEM early in life, and shows that the effects of such experiences can be long lasting.[176] For example, Nobel laureates, when asked what had led them to pursue science, frequently cited experiences such as experiments done out of school (such as with home chemistry sets), books that had captured their imagination, and teacher who had inspired and guided their personal interests.[177] Teachers, families, and communities can all play a role in fostering children's experiences that spark interest in STEM. The large number of people in the scientific and engineering communities who are willing to devote time to aiding students, including the more than a million engineers and scientists in the United States over age 60 and the more than half a million doctoral students and postdoctoral trainees, could represent a great asset for such activities.[178]

While shared standards can help ensure that all students reach certain levels of knowledge and capability in STEM subjects, standards alone are not sufficient to provide what the Nation needs in STEM education. Students need opportunities to establish deeper engagement with and to learn science and mathematics in non-standard, personal, and team-oriented ways that extend beyond the curriculum and the classroom. This is especially vital for identifying and nurturing high achievers and future STEM innovators.

Out-of-Class and Extended Day Activities

Out-of-class and extended day activities in science, technology, engineering, and mathematics can take many forms and can serve students regardless of their level of achievement in a formal learning environment. Students could participate in school-based mathematics contests, science fairs, or robotics competitions. They could attend after-school programs in which they connect mathematics with household activities, build robots using Legos, tell stories with animation technology, or study wildlife. They could visit museums, zoos, and aquaria or work as interns in science centers and state parks.[179] They could enroll in weekend programs or summer camps on engineering, astronomy, or number theory. They could work as interns with STEM-oriented companies in their cities and towns, attend lectures by STEM practitioners, or be mentored or tutored by people in their communities who work in STEM fields. They could use design labs that give them access to sophisticated fabrication laboratories. Evaluations of such programs rated as high quality show that they are associated with increases in student achievement and other positive outcomes.[180]

Opportunities to learn STEM outside of school are especially important for members of groups underrepresented in science and engineering, including girls, African-Americans, and Hispanics. As early as elementary school and middle school, many students from these groups begin to think that they will not or cannot excel at STEM. These messages sap the natural interest and lower the performance of groups underrepresented in STEM fields. Out of class activities can counter these messages by showing students that they can succeed at STEM and that STEM is not reserved for certain types of students. Also, they can connect students with role models and mentors who take an interest in their success in those fields and believe in their promise. In addition, group-based STEM activities can play an important role in forming peer groups interested in STEM and in changing the culture of STEM to attract students with diverse backgrounds, interests, personalities, and learning styles. In this way, out-of-class activities can build interest and persistence in STEM subjects for girls and the members of minority groups.[181]

Out-of-class activities also offer one way for students who are at risk of failing classes or dropping out of school altogether to get back on track. These activities can be more personalized than in-class instruction, which means that they can identify the point of connection between a STEM field and an individual student and build on that connection. Many struggling students who later achieve success in STEM professions gain their first real experience with the fields in an after-school activity.[182]

Out-of-class activities are of equal importance to high-achieving students. Such students can be at risk of abandoning their interest in STEM subjects, including failing classes or even dropping out of school, because they are bored by their coursework and not inspired by their teachers. Out-of-class activities give these students an outlet for their abilities in STEM subjects and new challenges. Activities that have interested many high-achieving students have diminished in recent years, and teachers report that they have less time to spend on these students as they devote increasing attention to low-performing students to meet accountability requirements.[183] Yet these out-of-class activities can be an important way to motivate and inspire the next generation of world-leading researchers and innovators. However, Federal funding for such programs has dwindled, and one of the last remaining Federal initiatives along these lines, the Javits Program, is probably about to lose the ability to focus on nurturing the gifted and talented.

There are a variety of approaches to STEM-focused activities:

After-school programs. Millions of children already participate in out-of-school programs. Most of these programs have no particular connection to STEM but could be adapted to include such connections. For example, the Twenty-First Century Community Learning Centers are after-school programs funded by the Federal Government under Title IV of the Elementary and Secondary Education Act. Many of these centers are in communities with low historical participation in STEM fields, and they typically have focused on youth development rather than STEM subjects. These community learning centers could provide enriching experiences in STEM if instructors were provided with engaging materials and the preparation and guidance to use those materials.

STEM contests. Out-of-class STEM contests that reward creativity and problem solving can be particularly powerful experiences for students (see Box 7-1). They involve students deeply in a subject and introduce them to a community of like-minded peers. Examples include the Siemens Competition in Math, Science and Technology, Mathcounts, the Intel Science Talent Search and International Science and Engineering Fair, the FIRST Robotics Competition (described below), and many exclusively local and regional competitions. More students would be able to participate in these competitions if there were more such contests and students were properly supported. Students in under-resourced communities may need funding for entry fees, materials, and travel expenses to participate in those programs. Teachers who elect to engage and support these students often need training, materials, and supplemental pay.

Designing and building. Out-of-class and advanced day activities also provide excellent opportunities for extended projects based on inquiry, construction, and discovery. These projects can combine digital and physical resources in ways that interest all students, not just those with a pre-existing interest in STEM. For example, the Playfully Inventing and Exploring (PIE) project uses hands-on, inquiry-based investigations to enable students to create art, automated devices, and narratives. Fab Labs (see Box 7-2) provide students with equipment and materials that they can use to invent, design, and construct an essentially limitless range of devices. More recently, the Maker's Faire has spawned a movement to have young people in after-school programs devise and carry out their own do-it-yourself projects. These opportunities spark interest in STEM while also providing students with the skills, thinking capabilities, and background knowledge they need to excel in STEM subjects. Box 7-3 describes two more programs designed to instill STEM knowledge and skills in students traditionally under-represented in STEM fields.

Summer programs. Many of today's STEM professionals were inspired to choose careers in STEM by programs sponsored by the Federal Government. In the 1960s and 1970s, for example, the National Science Foundation supported the Student Science Training Program (SSTP) for high-ability secondary school students.[184] These students participated in in-depth classes or research projects during the summer, typically between their junior and senior years. The objective of the program was to build the interest of science-oriented high school students through direct experience with college-level instruction and research. The Federal Government has eliminated many of its programs for highly talented students.

Federal Support for Out-of-Class and Extended Day Activities

The Federal Government can and should play a significant role in expanding opportunities for high-quality STEM-focused out-of-class activities. This will require a wide-ranging and thriving ecosystem of programs that respond to the needs of children as well as communities. Kids need more time to learn, and they need to learn in different ways – including through both extended day and afterschool programs. Moreover, different communities have different needs.

Box 7-1. STEM Contests

Contests and demonstrations involving science, technology, engineering, and mathematics can be tremendously rewarding for students who compete in them, regardless of where they finish in an event. From the Science Bowl to the Intel Talent Search, millions of students each year gain experience with the methods and subject matter of STEM through demonstrations and contests. Contests involving technology give students an opportunity to build something that exercises their creativity and reasoning. For example, the FIRST Robotics Competition challenges high school students from around the world to solve a particular engineering problem each year. Teams are given a standard set of parts and game details in January, and they then have six weeks to construct a robot that performs specific tasks. Volunteer mentors, including engineers and technology professionals, provide guidance. In 2010 the program will reach nearly 48,000 students from throughout the United States as well as Brazil, Canada, the United Kingdom, Mexico, Chile, Germany, Israel, Turkey, Australia, and the Netherlands.

Math competitions also engage large numbers of students in the United States in STEM. Each year more than 250,000 middle school students use materials prepared by Mathcounts. Through a series of team and individual competitions at the school, district, and state level, 224 students are selected from 56 states and territories to participate in the Mathcounts national competition. In high school, many students compete in math clubs and in the series of competitions sponsored by the Mathematical Association of America. These competitions result in the selection of the six-person U.S. team that participates in the International Mathematical Olympiad each year. The U.S. team consistently finishes near the top of the competition.

Many students who become interested in mathematics through contests look outside their schools for opportunities to engage their intellectual passions and find communities of similar students. Over the last seven years, more than 65,000 middle and high school students interested in mathematics have gathered at a website sponsored by the Art of Problem Solving (www.artofproblemsolving.com). The website also provides a wide variety of free resources used by high-performing middle and high school mathematics students around the world, including an adaptive learning system, videos, and a community-built repository of problems and solutions. In addition, the Art of Problem Solving organization develops textbooks and online classes that use creativity and problem solving to teach the middle school and high school mathematics curriculum.

Federal agencies currently support out-of-class programs with STEM activities, but these efforts are often *ad hoc* rather than part of a coherent strategy. Some of the programs are successful but are not coordinated with other programs. Many of the programs are not built on solid educational expertise.

Box 7-2. Fab Lab

The Fab Lab program, run by the Massachusetts Institute of Technology's Center for Bits and Atoms, was designed for informal, peer-to-peer technical training, but innovative educators are harnessing its potential to engage and inspire children in STEM subjects. Fab Labs offer the tools needed to conceive, design, and fabricate a wide range of objects—from a communications circuit board to a small building. The labs house computers, machinery, and raw materials with which participants can build a wide array of inventions. For example, students have devised solar-powered cooking gear, replacement parts for out-of-date machines, mobile refrigeration units, wireless data networks, and a toothbrush with a built-in MP3 player. The labs aim to spark creativity and offer users the opportunity to learn by designing and creating objects of interest and importance.

Since MIT constructed the first Fab Lab in 2002, communities worldwide have taken advantage of them. More than 40 labs currently exist in countries around the world including, Afghanistan, Colombia, Norway and the United States. Fab Labs around the world can contact one another through an online network to exchange ideas and troubleshoot challenges. Students can tap into knowledge that they do not yet possess—and help others brainstorm solutions to demanding problems. Students can acquire many STEM capabilities through the Fab Lab program: They apply mathematical and geometrical knowledge, learn key engineering and design principles, evaluate materials and chemical properties, and gain proficiency in fabrication, electronics, circuit design, and programming.

A few U.S. schools have begun to incorporate Fab Labs into their curricula. In East Cleveland, Ohio, tenth graders at the MC^2STEM High School have fabricated solar-powered iPod chargers and light fixtures. Because their school sits on a STEM industry corporate campus, engineers and technicians help students and teachers with projects and test potential solutions to their own real-world problems in the Fab Lab, thus exposing students to STEM in the real world. In the United States a few thousand children are currently benefiting from formal and informal Fab Lab programs in science museums and other locations as well as their schools.

In addition, the Federal Government supports many programs that currently lack STEM-focused components, but it could reorganize those initiatives into a coordinated program that incorporates STEM. Most prominent are the 21st Century Community Learning Centers, mentioned earlier in this chapter, which serve more than 1.5 million students through after-school programs at nearly 10,000 centers.[185] Funding for STEM activities through these centers in the 25 states that report data for individual programs is estimated to be $156 million and would likely be much higher if programs from all states were included.[186]

These programs could provide many more opportunities in STEM if instructors were provided with high-quality materials and professional development. Given the nature of these programs, online professional development could be a particularly promising route. Out-of-class and extended day programs also offer fertile ground for partnerships between government and the private sector.While we know that individual experiences are critical to attracting many students to STEM, we lack systematic knowledge about which types of programs serve best to inspire students to pursue STEM, and which qualities of successful programs are important to replicate. Even in the absence of such information, it is important to vigorously support programs that provide individual STEM-focused experiences. But it is important that we study these programs over time and learn from our investments.

BOX 7-3. STEM FOR ALL

Programs that seek to engage underrepresented groups of students in STEM, including minorities and girls, have demonstrated that out-of-school and extended day activities can spark interest and provide needed encouragement and mentoring in STEM. Two examples, Project Exploration and Expanding Your Horizons, are described here:

Project Exploration, a nonprofit science education organization, offers experiences with science to traditionally underserved communities, especially minority youth and girls. Its free-of-charge programs engage more than 250 Chicago public school students each year. Participation does not require previous academic success; rather, open-minded, curious children are invited to explore science with like-minded peers under the guidance of adults who serve as role models as well as instructors.

Project Exploration strives to connect young people with working scientists outside the conventional classroom. For example, its Dinosaur Giants program trains public high school students to serve as docents at the Museum of Science and Industry, the Garfield Park Conservatory, and The Field Museum. The enterprise teaches students facts and history about particular exhibits and instructs the young docents to communicate with the public as they lead tours, answer questions, and facilitate educational programs. A different Project Exploration venture, the Junior Paleontologist program, immerses minority youth and girls in geology, anatomy, and paleontology – and then it sends them into the field to search for fossils alongside scientists. After the students return from this three-week summer program, they receive mentoring and leadership-development opportunities until their high school graduation.

Expanding Your Horizons, a nonprofit network of scientists, mathematicians, educators, parents, community and business leaders, and government officials, provides mentoring, encouragement, and support to girls to pursue STEM subjects, excel in them, and eventually pursue STEM careers. In 2010 the National Science Board awarded its prestigious Public Service Award to Expanding Your Horizons in recognition of this work. The network holds more than 80 conferences annually in 33 states and in Europe and Asia, which feature workshops to conduct hands-on STEM projects and experiments for girls to participate in with the help of STEM mentors. Since 1976, more than 775,000 girls have participated in these conferences, examining rocks and fish, building machines, dismantling computers, and solving problems. The network also connects girls with STEM mentors and trains educators to better support girls in STEM subjects.

Recommendation 7-1. Inspiring Students through Direct Experiences

The Department of Education and National Science Foundation, together with the other Federal science mission agencies, should develop a coordinated initiative – which we will call INSPIRE 187 – to provide students with high-quality opportunities for individualized, transforming experiences with STEM subjects.

INSPIRE should provide vigorous support for high-quality programs that can inspire students about STEM subjects, including:

- **High-quality STEM activities in after-school and extended day programs, together with support for programs to train providers and develop high-quality instructional materials.**
- **Local, statewide, and national STEM contests, including increasing the number of such contests, extending participation to more schools and communities, and supporting programs that prepare instructors and participants.**
- **Programs and facilities that let students design and create their own devices and machines such as through access to sophisticated fabrication capabilities.**
- **Summer programs for high school students, such as a revival of the National Science Foundation's Student Science Training Program.**

Special attention should be paid to programs that focus on engaging and addressing the needs of girls, minorities underrepresented in STEM fields, and students with disabilities. Federal agencies should maintain an inventory of such programs, assess the effectiveness of each program with respect to preparation and inspiration, compare results, and report regularly on these efforts. Schools should look to these out-of-school programs for ideas that might be brought into the classroom, and should consider recognizing student experiences in these programs where appropriate with academic credit.

Importantly, the agencies should also engage visionary corporations and foundations in developing and supporting programs under INSPIRE.

The science mission agencies currently support many out-of-school activities that could be part of this program, but they are not well coordinated. In addition, the Department of Education funds support many out-of-school and after-school programs, but these are often part of block grants and thus not readily targeted for STEM. The INSPIRE program could be supported by coordinated funding from the science mission agencies, together with targeting a portion of the Department of Education's 21st Century Learning Centers grants toward STEM (although this may require new authorities). Through such sources, it should be possible to devote $500 million annually in existing Federal funding to such initiatives.

We also note that the corporate and philanthropic sectors have played and can continue to play a significant role in supporting STEM-focused programs that inspire students. We urge that the Federal strategy in implementing this recommendation be complemented by and coordinated with a more substantial contribution of private and philanthropic groups, such as Change the Equation.

Advanced Classes

Opportunities for accelerated learning and new challenges within school also can give students a personal connection to STEM subjects. Advanced classes are one way to achieve this. These classes can exist at any educational level. "Pull-out" sessions in elementary school and middle schools can give students who want to go beyond the standard curriculum a way to explore their interests. Advanced courses at the high school level, as well as access to college courses, Advanced Placement (AP) courses, the International Baccalaureate (IB) program, and dual enrollment classes in high school, can offer students access to rigorous STEM study and high levels of achievement.

We focus on three opportunities at the high school level:

1. **Advanced Placement courses**. AP courses provide an important and meaningful opportunity to achieve an objective high standard that is nationally recognized. Student enrollment in these programs is one of the bright spots of STEM education in recent years.[188] More than 250,000 students passed a mathematics AP exam in 2008, compared with just over 50,000 in 1990. More than 200,000 students in public and private schools passed a science AP exam in 2008, compared with fewer than 50,000 in 1990. More students of both sexes and all minority groups took AP tests in these subjects in 2008 than in 1997.

   Many more students appear to be capable of performing AP work. Based on PSAT scores, there are 633,000 additional high school students in the United States who would be predicted to be able to pass AP Calculus, if given the opportunity with a trained teacher. Notably, only about 7,000 African-American and Hispanic students currently pass AP Calculus, but 123,000 would be predicted to be able to pass. Longitudinal studies have shown that successfully passing AP exams is correlated with successful outcomes in college. And specific programs that combine increased AP enrollment (especially for underrepresented students) with significant teacher professional development and extra time on task for students show higher four-year-college graduation rates among African-American and Hispanic students who participated in the program than among students who did not. [189] Such studies are difficult to interpret, because correlation does not imply causation. Still, there is reason to believe that participation in courses with high academic intensity, such as AP classes, is a positive experience in preparing students for later academic challenges.

   There are understandable concerns about the AP program: AP classes can cover a large number of topics at such a rapid pace that students and teachers can both feel rushed to cover the material, resulting in a superficial rather than robust grasp of important concepts. AP courses and AP exams generally do not promote inquiry-based science, which is critical for inspiring students to pursue STEM careers. Importantly, the College Board is engaged in work to change the AP curriculum and exams to address these issues. The Advanced Placement Training and Initiative Program (APTIP) is also attempting to address these issues. [190] We strongly endorse such changes, which ideally will have a ripple effect on the rest of the high school curriculum.

Box 7-4. Advanced Placement Training and Incentive Program

The Advanced Placement Training and Incentive Program (APTIP) aims to increase the number and diversity of students taking and passing College Board Advanced Placement courses and exams in mathematics, science, and English, with the goal of elevating participating schools' expectations for their students and transforming schools' cultures toward college-readiness for all.[192] The program, administered by the National Math and Science Initiative, Inc., was highlighted in the 2005 report *Rising Above the Gathering Storm* published by The National Academies of Science.

APTIP provides intensive teacher preparation, teacher and student support, open and encouraged enrollment in AP courses, financial incentives for students and teachers based on academic achievement, specific and individualized annual achievement goals, and robust data collection to ensure accountability at all levels. It cultivates lead teachers who mentor their colleagues, including those who teach pre-AP students and are thus best positioned to expand the number and diversity of students in the pipeline for future AP classes. The program includes recruitment, counseling, tutoring, and exam-preparation components designed to educate students and their families about AP programs, boost students' confidence, and enhance their chances of success.

APTIP has had considerable success in increasing the number of students taking and passing AP mathematics and science exams, especially for traditionally underrepresented students, including girls, African Americans, and Hispanics. APTIP is now working in 227 participating high schools across six states. Among the 65 schools that have participated in the program for the past two years, APTIP produced a 98 percent increase in AP mathematics, science, and English (MSE) exams passed; a 155 percent increase in AP MSE exams passed by African-American and Hispanic students; and a 116 percent increase in AP exams passed by female students in mathematics and science.

Some programs have been developed in recent years for training teachers to teach AP courses and for providing incentives to both teachers and students.[191] An example is discussed in Box 7-4.

The number of US public high school students who pass at least one AP exam in math and science (about 220,000 and 190,000, respectively) represents 5-6% of the total number of students in the senior year,[193] while the numbers for minority students (about 18,000 and 14,000) represent less than 2% of the total seniors in public schools for this group. In our recommendation below, we propose that the Federal Government should set a goal of doubling the overall numbers and tripling the numbers for minority students.

2. **International Baccalaureate**. Another option for students to take advanced classes is through school programs certified through the International Baccalaureate Organization (IBO). Currently, more than 600 schools in the United States offer the IB Diploma Program, a high school course of study with rigorous academic standards. Among other requirements, students must take four years of a foreign

language, complete a Theory of Knowledge class, and take six challenging examinations to receive an IBO diploma.

3. **College and online courses**. A third option for students seeking to take advanced STEM classes is enrollment in courses offered by nearby colleges and in online courses. This may be particularly important for students in low-income schools with limited course offerings. For example, a program of certified online courses could be offered to students who complete the course of study in mathematics, technology, or sciences offered at their school and want to advance to higher-level study.
   Such courses differ in quality both in terms of the material taught and the effectiveness of instructors. Greater evaluation of these dual-enrollment options would indicate whether they work well for students and how best to expand these opportunities.

We urge the Department of Education to analyze the funding that would be required to achieve these goals.

The administration requested $100 million in its FY11 budget, which the ESEA reauthorization proposal would allocate for the college pathways and accelerated learning program. The budget item would be for advanced courses in all subjects; the portion to be dedicated toward STEM courses has not been specified. Significantly more resources will likely be required to implement this recommendation.

RECOMMENDATION 7-2. ADVANCED COURSES

The Federal Government should take steps to increase substantially the number of students who excel in challenging STEM subjects by taking courses that go beyond the standard curriculum. It should:

- **Set a goal to double the number of students who take and pass Advanced Placement courses in STEM subjects and triple the number of minorities underrepresented in STEM who take and pass these courses, within the next 5 years. To help achieve this goal, it should provide support for programs with proven ability to support such increases.**
- **Support programs to increase the number of students taking and passing other advanced courses, such as those offered through the International Baccalaureate program and dual enrollment college classes.**
- **Provide financial support to high school students who have completed the STEM requirements at their high school and wish to take additional courses not available at their schools by offering dual enrollment opportunities that allow them to attend a local college or take high-quality online courses, provided that they are set up and evaluated properly.**

VIII. Schools and School Systems

Chapter Summary

All schools need connections with the outside world of STEM to demonstrate to their students the vibrancy and potential rewards of STEM-related professions. The United States needs many more schools at all levels that are focused on STEM subjects, especially in high-poverty neighborhoods. These schools can serve as valuable testing grounds for innovative approaches to STEM teaching and learning, and they can provide focused experiences in STEM that prepare and inspire students for lifelong study. Even though all schools cannot and should not be deeply STEM-focused, all schools should have partners in business, academia, or the nonprofit sector that can bring STEM subjects to life for students and help students and teachers understand the value of STEM knowledge in the real world. The Federal Government should support the creation of at least 1,000 STEM-focused schools over the next decade, 200 at the high school level and 800 at the middle and elementary school levels. In addition, the Federal Government should use its leadership and voice to help ensure that every school has access to STEM expertise by supporting partnerships for schools. Finally, the Federal Government should invest in the development of STEM education leaders at the school, district, and state levels.

Introduction

The dynamic nature of science, technology, engineering, and mathematics – where new advances are constantly expanding our knowledge of the physical, biological, and social world – has enormous implications for STEM education. By making abundantly clear to students how many discoveries remain to be made and the role young people can play in solving important problems, schools and school systems can excite and motivate students to learn science and mathematics and to pursue careers in STEM fields.

To seize these opportunities, teachers and schools must connect students with the dynamic world of STEM. In turn, this requires that teachers, schools, and school systems have a deep understanding of STEM activities in American society and business and that they maintain direct connections to appropriate STEM expertise.

In this section, we discuss several solutions to this challenge, including creating many more specialized STEM-focused schools and forging connections between all schools and the STEM community.

STEM-Focused Schools

The United States currently has about 100 public high schools[194] that make ambitious efforts to focus on STEM subjects. These schools create groups of grade school peers with common interest in STEM, expose students to advanced STEM content, provide students with opportunities for exploration and discovery, connect students to real-world STEM workplaces, and provide role models to students in STEM professions. These schools include charter schools, magnet schools, pull-out programs, and boarding schools. They currently enroll about 47,000 students, most at the high school level.

Among the first STEM-focused public high schools were Stuyvesant High School (founded in 1904) and the Bronx High School of Science (founded in 1938) in New York City. They have trained generations of future leaders in STEM. More recent prominent examples and High Tech High (see Box 8-1) in San Diego (founded in 2000), the Illinois Mathematics and Science Academy (founded in 1985) in Aurora, Illinois (see Box 8-2), Thomas Jefferson High School for Science and Technology (founded in 1985) in Alexandria, Virginia (see Box 8-3). Other STEM-focused schools include North Springs Charter High School in Atlanta, the New Orleans Charter Science and Math High School, the Delta High School in Washington State, and Metro Early College High School in Ohio. The number of STEM-focused schools has been growing somewhat in recent years. Some STEM-focused schools have a required entrance examination while others have open enrollment.

Despite this growth, highly-STEM-focused schools remain a rarity in the United States, enrolling fewer than 1 student in 1,000. Only 30 states have rigorous schools or programs that recognize and cultivate STEM talent. Some states, such as Georgia, New York, Michigan, and Virginia, have a high concentration of these programs. Most STEM-focused schools are singular creations, with few attempts to scale successful schools. Few programs are directly targeted at underrepresented groups. Furthermore, few STEM-focused schools are found at the elementary and middle school levels, even though studies show that student interest – or disinterest – in STEM can solidify by middle school. There are a few examples of elementary schools with rich STEM connections, such as the NASA Explorer Schools, but we do not know enough to know which models work best.

We need many more examples of STEM-focused schools at all levels, especially those that serve minority communities. Given the success of the open enrollment Knowledge is Power Program (KIPP) Schools in scaling up highly effective middle schools in minority communities and in improving mathematics achievement in this setting,[195] it would be especially exciting to see KIPP-like models of STEM-focused schools. Such schools could change attitudes toward STEM and engage much larger numbers of students from underrepresented groups than is the case today.

STEM-focused elementary schools could also provide a unique opportunity to better connect science learning and literacy. Currently, reading and science are generally taught as distinct subject areas, and the potential for synergies between the two areas of learning are often overlooked. Science texts can enrich and enliven the process of acquiring literacy for many students by tapping into content that is current and dynamic. Similarly, literacy can enrich the learning of STEM by helping students formulate hypotheses, make sound arguments, keep journals of observation, and access scientific information from a variety of sources.[196] Looking for ways that the teaching of reading and writing and the teaching of science can overlap and be complementary could increase the amount of classroom time devoted to each of the subjects, empowering students to access STEM subjects through multiple entry points. STEM-focused schools at the primary level could provide a testing ground for this approach and for related instructional materials.

It is difficult to measure the precise impact of STEM-focused schools through formal trials or analyses for two reasons. First, STEM-focused schools take many different approaches. Some emphasize traditional academic preparation while others, like High Tech High, connect high-level academics to preparation for more specific careers. Second, the students who choose to attend these schools are not a random sample, and it is hard to identify a suitable control group. The available quantitative studies support the notion that STEM-focused schools produce students who take STEM majors in colleges at disproportionately high rates.[197] And abundant observational data make a compelling case that STEM-focused schools have a major impact on their students. In addition to their direct impact on students, STEM-focused schools are laboratories for experimenting with creative approaches to STEM education that can have a broader impact on U.S. education through the dissemination of ideas, programs, and materials that can benefit all students.

BOX 8-1. HIGH TECH HIGH

In 2000 a group of civic and high-tech industry leaders came together to launch a new public charter high school in San Diego that would train students from the city's diverse ethnic, racial, and socioeconomic communities to excel in STEM, and in particular to prepare to become high-tech industries' workforce of the future. Since 2000 the program has expanded to include five high schools, three middle schools, and one elementary school in Southern California, enrolling about 3,500 students.

High Tech High features up-to-date technology resources and laboratories available to students. It has classrooms modeled after high-tech workplaces, where students gather in small groups to collaborate on problems. The students do their own innovative, hands-on projects that range from building human- powered submarines to studying the ecology of San Diego Bay to making documentaries about how medicine moves through the human body. High Tech High fosters connections between its students and high-tech workplaces by requiring academic internships at local businesses and research organizations, where students learn such skills as web design, video editing, and mobile technology analysis. It also gives students opportunities to shadow STEM practitioners in their workplaces throughout high school and to find mentors and role models in fields of interest.

To date, all graduates of High Tech High have been accepted to college. More than 30 percent of High Tech High graduates pursue STEM fields in college, compared to national rates around 17 percent. The schools also address disparities in STEM achievement: 35 percent of their graduates are first-generation college students, and their African American students outperform their African American peers in California in science and mathematics test scores and in taking advanced courses in mathematics, chemistry, and physics.

Given the tiny number of STEM-focused schools in the United States, it is clear that the Nation would benefit from having substantially more STEM-focused schools, with varied approaches and serving demographically varied groups. Whatever the optimal number of such schools for the Nation, we are far below this level.

There are many possible approaches to creating STEM-focused schools. They can have varied themes, instructional approaches, and partnerships. We need many more experiments with new approaches, as well as experiments in scaling up successful approaches.

Box 8-2. Illinois Mathematics and Science Academy

In 1985 the state of Illinois launched the Illinois Mathematics and Science Academy (IMSA), a STEM- focused, public residential school that enrolls 650 high school sophomores, juniors, and seniors. Since then, about 65 percent of its nearly 4,000 alumni have pursued degrees in STEM, and its female graduates earn STEM degrees at more than three times the national rate among women. In 2009, Intel named IMSA the nation's Star Innovator School for its consistent excellence in science education. While most of the school's operating budget comes from the state, it also brings in foundation, local, and Federal grants.

IMSA provides its students, selected by a rigorous application process from among the state's high school ninth graders, with rigorous and challenging STEM courses, laboratory experiences, and opportunities to solve problems, design technologies, and pursue scientific inquiry outside of classrooms. About half of IMSA faculty hold doctoral degrees.

IMSA students have the opportunity to discover STEM knowledge on their own and to direct their own learning. Their initial task when they enroll in the academy is to work with teachers and mentors to design a personalized learning program that includes the courses and activities they will pursue. In the Student Inquiry and Research Program, a year-long independent study course, IMSA students are paired with advisors who are STEM professionals in academia or industry. The students spend one day of each week for the academic year pursuing their own investigation, which they ultimately publish and present to the school community. Through this program, IMSA students have studied the magnetic qualities of nanoparticles, infant mortality rates, mercury contamination in Lake Michigan, and models for predicting influenza outbreaks. Their projects answer questions in neuroscience, economics, and particle physics, and many win national and international awards.

The creation of STEM-focused schools is the responsibility of state and local authorities, but the Federal Government should vigorously encourage and support such efforts by providing funding and technical support. This is particularly important because creating a high-quality STEM-focused school can be more expensive than creating a typical school, owing to special programmatic and infrastructural needs. Despite their somewhat higher cost, high-quality STEM-focused schools are a superb investment in the Nation's future.

We estimate that the Department of Education would need to provide financial support totaling at least $10 million per high school and $2 million per middle and elementary school. This support should focus on planning, professional development, materials, laboratories, technology, and equipment. Over the next decade, the total cost of the program would thus be approximately $360 million per year.

Some resources could come from existing Federal programs, but additional Federal funds will likely be required. It may also be appropriate for States to provide some matching funds. In addition, the Federal Government should actively engage the private and philanthropic sectors in the creation of new STEM- focused schools.

BOX 8-3. THOMAS JEFFERSON HIGH SCHOOL FOR SCIENCE AND TECHNOLOGY

In 1985, representatives from the Fairfax County Public School System and local businesses teamed up to establish the Thomas Jefferson High School for Science and Technology. The school is administered by the Fairfax County Public School System and also serves as the Governor's School for Science and Technology in Northern Virginia. It enrolls approximately 1,800 students in grades 9-12 through a highly competitive selection process based on students' aptitude for and interest in mathematics and science as well as their intellectual curiosity, motivation, and integrity.

Consistently ranked as one of the best high schools in the United States, Thomas Jefferson provides its students with rigorous classroom experiences and opportunities to use the school's 13 state-of-the-art research laboratories. Students are paired with mentors from universities, industry, and government and participate in research projects in astrophysics, neuroscience, and microelectronics. Each year, Thomas Jefferson students are among the finalists in the prestigious Intel Science Talent Search. Students publish a research journal, called TEKNOS, where they publish their peers' research projects.

During the 2008-09 school year, more than 99 percent of Thomas Jefferson students took Advanced Placement courses and more than 99 percent received a score of 3 or higher on the AP exam they took. One in every three seniors earns National Merit Semifinalist recognition, placing them in the top 0.5 percent of all students nationally.

RECOMMENDATION 8-1. CREATE 1,000 NEW STEM-FOCUSED SCHOOLS

The Department of Education, in collaboration with the National Science Foundation, should support the creation by the states of 200 new highly-STEM-focused schools at the high school level and 800 STEM-focused elementary and middle schools over the next decade.

These schools would serve the Nation both through the students they produce and by being testing grounds for approaches to STEM-focused education, providing scalable models for STEM-focused schools. Particular attention should be paid to creating new types of STEM-focused schools, including those serving minority and high-poverty communities; these schools could play a special role in closing the achievement gap in STEM fields.

The Department of Education and the National Science Foundation should develop a joint plan for accomplishing this goal, through funding and technical support.

To provide resources for state and local authorities seeking to create new schools, the agencies also should commission the National Research Council to study current and new options for STEM-focused schools.

We note that 1,000 new STEM-focused schools would constitute only about 1 percent of the 98,000 public schools in the Nation. There is thus no significant risk that these new schools will drain away a significant proportion of the high-quality STEM teachers from other public schools. A more likely outcome is that such schools may attract new teachers into the public schools, perhaps including teachers from private schools.

Creating Bridges from Schools to STEM Expertise

Most schools in the country will not be and should not be STEM-focused schools, but all schools should have a direct conduit to STEM expertise. Thriving connections to STEM expertise can bring to life the material in textbooks. Such connections can bring teachers into contact with the excitement at the cutting edge of science. They can link students with role models who work in STEM careers, counteracting the stereotypes that keep many students from engaging in STEM subjects. And schools with STEM expertise can more readily take advantage of opportunities to turn their communities into laboratories for the learning of STEM.

A variety of programs attempt to bridge the gaps between public schools and the STEM professional community, but not all such programs provide teachers and schools with resources that are useful in their classrooms. Nonetheless, several programs demonstrate the potential for such connections to benefit K12 schools. For example, Teachers Institutes, which began in 1978 in New Haven and have since expanded to cities across the country, pair universities and school districts, allowing teachers to identify the topics on which they would like to collaborate. University professors then guide these teachers through inquiry- based learning in a STEM subject area. More recently, National Lab Day (described in Box 8-4) has linked STEM professionals with K-12 teachers and schools for projects and lab experiments using a technology- based matching service similar to that used for online dating. The United States is home to more than one million STEM professionals over age 60, many of whom have retired and could constitute a vast cadre of volunteers for K-12 schools. It is important that we find way to harness these sources of partnership and expertise in a committed, sustained way relevant to K-12 teachers and students.

Elementary Schools. Elementary teachers face constraints in teaching STEM that include insufficient content knowledge and lack of confidence, lack of materials and facilities, and lack of support from their schools.[198] We must address these gaps if we hope to engage students at a young age in STEM. Most elementary teachers are generalists and are not trained specifically in STEM fields; the subjects can be a source of anxiety and lead to avoidance. While it would not be practical for all elementary teachers to become STEM experts, primary schools can and should have at least one expert teacher in science and one expert teacher in mathematics. Such highly knowledgeable resident STEM experts can serve as leaders in their schools, providing expertise and mentoring to their colleagues on how to illustrate and animate their subject areas with STEM content. Elementary schools also need critical materials, labs, and resources to provide rich opportunities for STEM learning. It should be noted that in some parts of the country, middle schools and teachers face the same constraints as do elementary schools and teachers.

Box 8-4. National Lab Day

Contrary to its name, National Lab Day (NLD) is not an annual commemorative event but a network of scientists, engineers, mathematicians, and educators launched in November 2009. NLD is supported by a coalition of STEM organizations, philanthropies, schools, teachers, and the White House. National Lab Day is not a one-day event but rather a network of STEM professionals and teachers who want to provide rich STEM experiences to students year-round. More than 200 partner organizations pledged to bring STEM learning opportunities to students in grades 6 through 12, in what the grassroots network's leadership dubbed "a national barn-raising for hands-on learning." Through a website (www.nationallabday.org) that has a matching program similar to those used on online dating websites, the initiative provides a venue for teachers who are seeking lab experiments, curriculum ideas, or outof-classroom experience to find STEM professionals who can help. Scientists, technologists, mathematicians, and engineers who register on the site are paired with teachers and schools to create experiments, find web resources or technological tools, design lesson plans, or provide laboratory or field experiences. Foundations and companies donate to support projects that require financial assistance. People who are not teachers or scientists can also help broaden the reach of the program by disseminating the requests and projects to their social networks and blog and Twitter audiences. The top scientists, educators, and donors are recognized on the website to encourage more participation.

The goal of National Lab Day is to carry out 10,000 hands-on learning projects reaching 1 million students by the end of the year. Currently, there are projects underway in all 50 states inside and outside of classrooms. More than 1,200 schools are participating, and about 4,000 teachers and about 4,000 scientists have registered on the site. On May 12, 2010, a day of recognition was held for National Lab Day.

Recommendation 8-2. Connecting Schools to STEM

The Department of Education, working with the National Science Foundation, should help ensure that all schools and school systems have access to relevant STEM-expertise by setting a national goal, and developing and supporting programs, to ensure that:

(1) every middle and high school has a STEM partner organization, such as a research organization, college or university, or a STEM-based nonprofit or company, to provide direct connections for teachers and students to STEM practitioners;

(2) every elementary and middle school has at least two highly knowledgeable STEM expert teacher-coaches, one in mathematics and one in science, who can serve as resources for the content and teaching of mathematics and science;

(3) all schools can readily draw on the expertise of volunteer STEM professionals, especially retired scientists and engineers.

Middle and High Schools. High schools, and most of the Nation's middle schools, typically have teachers with some STEM expertise, but these teachers often lack meaningful connections to the STEM professional community. Such connections can help schools, teachers, and students explore cutting-edge content and real-world applications. Every middle school and high school should have a partner in a STEM field, such as a research organization, college, university, museum, zoo, aquaria, or company, that can bring STEM subjects to life for students and help teachers and students learn about STEM in the workplace.

The NSF and the Department of Education can begin work toward these goals with existing resources, although some additional resources may ultimately be required. In addition, the private and philanthropic sectors can and should play a critical role in the work of connecting schools to STEM and engaging STEM professionals. The Federal Government should systematically engage these sectors, potentially through the leadership of such consortia as the recently-formed group, Change the Equation.

Ensuring that Education Leaders are Knowledgeable about STEM Education

Successful STEM education requires the support of school principals and superintendents, yet these leaders often lack an understanding of STEM fields or STEM education. If school leaders had greater awareness of these subjects, they would be more likely and more able to cultivate rich STEM learning experiences and expertise in their schools. Furthermore, barriers remain in connecting research on STEM teaching, learning, and leadership with practice. Developing and implementing mechanisms to better connect the research and evaluation world with those charged with leading systemic change would be an important reform.

The Federal Government is uniquely positioned to use agency resources, research capacity, and its convening function to help expand STEM knowledge and exposure among school and school district leaders. We suggest the following initiatives for consideration:

Research on leadership development. The NSF should focus research and development to identify the knowledge needed by school principals to effectively lead STEM instruction.

Scale up STEM-focused leadership development programs. The NSF and the Department of Education should invest in programs to prepare school leaders to lead STEM schools and to incorporate STEM-focused learning into existing leadership development programs (such as the Lenses on Learning Program funded by the NSF). A benchmark study that describes the STEM experiences of school leaders – both what is required by certification rules and what is true in practice – would be a good first step. An explicit focus on high school departments and department chairs (the locus of leadership in most high schools) would be important.

STEM leaders conference. The NSF, in conjunction with the Department of Education and the Council of Chief State School Officers, should create a series of conferences for state chiefs, state-level STEM leaders, and the superintendents and curriculum leaders of the 50 largest school districts. These conferences— which would ideally convene several times a year—would focus on (a) disseminating current research about effective STEM programs, (b) analyzing current implementation data from attending states and districts to collectively share progress, and (c) soliciting the field for key research questions that need to be answered.

Enhance technical assistance around STEM leadership. With its Race To The Top Program, the Department of Education has a unique opportunity to shape state and district system policies and programs governing STEM teaching, learning, and leadership for the foreseeable future. The Department should invest in systems and mechanisms to move from technical assistance focused on traditional compliance and accountability to a strategy focused on increasing capacity. Since Race to the Top includes STEM as a competitive priority, there should be a strong focus on STEM within this new technical assistance work.

STEM audits. The Department of Education and NSF should explore how to create an organization external to the Federal Government that could provide district-level audits of STEM programs, much in the way that companies like Cambridge Education provide this service for schools. These audits or inspections would provide objective data to district leaders about the quality and breadth of their STEM education programs and ratchet up conversation about how to improve them.

Recommendation 8-3. Connecting Education Leaders with Best Practices in STEM Education

The administration should help ensure that every education leader – including school principals, district leaders, and state superintendants – has knowledge of the unique issues and best practices in achieving excellent STEM education through programs such as those described above.

We believe that the NSF and the Department of Education can accomplish these goals through existing programs, although some additional resources may be required.

Appendix A: Experts Providing Input to PCAST

PCAST expresses its gratitude to the following individuals who provided input by attending meetings or by responding to requests for information:

Mark Atkinson
Chairman & CEO
Teachscape

Larry Berger
CEO

Al Cuoco
Senior Scientist and Director of the Center for Mathematics Education
Education Development Center

John Ewing

Wireless Generation

Jim Behnke
Chief Learning Officer
Pearson Education

Rodger Bybee
Chair of Science Forum and Science Experts Group
PISA

Margaret Cagle
Award-winning K-12 STEM Teacher

Michele Cahill
Vice-President
National Programs
Carnegie Corporation

Dan Caton
Executive Vice President
McGraw-Hill School Education

Karen Cator
Director, Office of Educational Technology
U.S. Dept. of Education

Richard Ingersoll
Professor of Education and Sociology
University of Pennsylvania

David Ireland
Award-winning K-12 STEM Teacher

Andy Isaacs
Co-Director, Center for Elementary Mathematics and Science Education
University of Chicago

Paul Jesukiewicz
Advanced Distributed Learning

President
Math for America

Joan Ferrini-Mundy
Acting Director, Education and Human Resources
National Science Foundation

Martin Gartzman
Assistant Vice Chancellor for High School Development
University of Illinois at Chicago

BJ Goergen
Chief of Staff
National Math and Science Initiative

Phillip Griffiths
Professor of Mathematics Emeritus
Former Director
Institute for Advanced Study

Toby Horn
Co-Director
Carnegie Academy for Science Education

Steve Robinson
Special Assistant
Domestic Policy Council

Larry Rosenstock
CEO
High Tech High

Brian Rowan
Burke A. Hinsdale Collegiate Professor in Education; Research Professor, Institute for Social Research
University of Michigan

Philip A. Schmidt

Initiative
Office of the Secretary of Defense,
Department of Defense

Michael C. Lach
Special Assistant
US Department of Education

Glenda Lappan
University Distinguished Professor
in the Division of Science and
Mathematics Education
Michigan State University

Camsie Matis
Award-winning K-12 STEM teacher

Andrew Porter
Dean, Graduate School of Education
University of Pennsylvania

Helen Quinn
Professor Emerita of Physics
Stanford University

Mary Ann Rankin
Dean, College of Natural Sciences
University of Texas at Austi

Steve Ritter
Co-founder and Chief Scientist
Carnegie Learning

Jo Anne Vasquez
Vice-President and Program
Director
Arizona Transition Years Teacher
and Curriculum Initiative

Chuck Vest
President
National Academy of Engineering

Gerry Wheeler
Emeritus Executive Director

Associate Provost
Academic Programs
Western Governors University

James Shelton
Assistant Deputy Secretary for Innovation
and Improvement
U.S. Department of Education

Joy Smith
Chief Development Officer
Florida Virtual School

Nancy Songer
Professor of Science Education and
Learning Technologies
University of Michigan

Jennifer Thompson
Award-winning K-12 STEM Teacher

Lisa Towne
Senior Consultant
Education First

Lee Umphrey
Vice President
Math for America

Suzanne Wilson
Chair, Department of Teacher Education
Michigan State University

Lee Zia
Director
NSF National Science, Mathematics,
Engineering and Technology Education
Digital Library

National Science Teachers
Association

Grover J. Whitehurst
Director of the Brown Center on
Education Policy
Brookings Institution

APPENDIX B. ACKNOWLEDGEMENTS

PCAST wishes to express gratitude to the following individuals who contributed in various ways to the preparation of this report:

TJ Augustine
Student Volunteer, PCAST

Jessie Dearo
Senior Policy Analyst
Office of Science and Technology Policy

Diane DiEuliis
Assistant Director for Life Sciences
Office of Science and Technology Policy

Pamela Flattau
Research Staff Member
Science and Technology Policy Institute

Daniel L. Goroff
Program Director
Alfred P. Sloan Foundation

Marta Kowalczyk
Adjunct Research Staff Member
Science and Technology Policy Institute

Thomas Kalil
Deputy Director for Policy
Office of Science and Technology
Policy
Senior Advisor for Science,
Technology, and Innovation
National Economic Council

Darius Singpurwalla
Adjunct Research Associate
Science and Technology Policy
Institute

Carl Wieman
Consultant
Office of Science and Technology
Policy

End Notes

[1] Claudia Golden and Lawrence F. Katz. (2010). *The Race Between Education and Technology*. Cambridge, MA: Harvard University Press.

[2] Organisation for Economic Co-operation and Development. (2000). *Science, Technology and Innovation in the New Economy*. Washington, DC: OECD.

[3] U.S. Council of Economic Advisors. (2000). *Economic Report to the President, 2000*. Washington, DC: U.S. Government Printing Office.

[4] Elhanan Helpman. (2004). *The Mystery of Economic Growth*. Cambridge, MA: Harvard University Press.

[5] Bureau of Labor Statistics. (2009). *Occupational Outlook Handbook, 2010-11, Bulletin 2800*. Washington, DC: U.S. Department of Labor. Accessible at http://www.bls.gov/oco/oco2003.htm.
[6] Patrick Gonzales, Trevor Williams, Leslie Jocelyn, Stephen Roey, David Kastberg, and Summer Brenwald. (2009). *Highlights from TIMSS 2007: Mathematics and Science Achievement of U.S. Fourth- and Eighth-Graders in an International Context*. Washington, DC: U.S. Department of Education.
[7] National Science Board. (2010). *Science and Engineering Indicators: 2010*. Arlington, VA: National Science Foundation. Accessible at http://www.nsf.gov/statistics
[8] Ibid, Chapter 2.
[9] Ibid, Chapter 1.
[10] B. Lindsay Lowell, Hal Salzman, Hamutal Bernstein, and Everett Henderson. (2009). *Steady as She Goes: Three Generations of Students through the Science and Engineering Pipeline,* paper presented at the Annual Meeting of the Association for Public Policy Analysis and Management, Washington, DC, November 5-7. Accessible at http://policy
[11] AAUW. (2010). *Why So Few? Women in Science, Technology, Engineering, and Mathematics*. By Catherine Hi l l, Ph.D. Christianne Corbett Andresse St. Rose, Ed.D. Washington, DC: AAUW.
[12] National Science Foundation. (2009). *Women, Minorities, and Persons with Disabilities in Science and Engineering: 2009*. Arlington, VA: National Science Foundation. Accessible at http://www.nsf.gov/statistics
[13] G. Ellison and A. Swanson. (2010). "The Gender Gap in Secondary School Mathematics at High Achievement Levels: Evidence from the American Mathematics Competitions." *Journal of Economic Perspectives* 24(2):109–28.
[14] National Science Board. (2010). *Science and Engineering Indicators: 2010*. Arlington, VA: National Science Foundation. Accessible at http://www.nsf.gov/statistics. On the NAEP for mathematics, the average fourth grade score rose from 213 to 240 between 1990 and 2007. For eighth graders, the average score rose from 263 to 281. The most recent NAEP results, however, show that student gains at the fourth grade level did not continue from 2007 to 2009.
[15] National Science Board. (2010). *Science and Engineering Indicators: 2010*. Arlington, VA: National Science Foundation. Accessible at http://www.nsf.gov/statistics. The average score gap between black and white fourth graders shrank from 32 to 26 scale points between 1990 and 2007, and the average gap decreased from 2000 to 2007 between black and white eighth graders after increasing between 1990 and 2000.
[16] National Science Board. (2010). *Science and Engineering Indicators: 2010*. Arlington, VA: National Science Foundation. Accessible at http://www.nsf.gov/statistics
[17] American Institute of Physics Statistical Research Center. (2010). *High School Physics Courses and Enrollment* (White Paper, August 2010), by Susan White and Casey Langer Tesfaye. Melville, NY: American Institute of Physics.
[18] Deborah D. Stine and Christine M. Matthews. (2009). *The U.S. Science and Technology Workforce*. Washington, DC: Congressional Research Service.
[19] National Research Council. (2005). *How Students Learn: History, Mathematics, and Science in the Classroom.* Washington, DC: National Academies Press.
[20] National Research Council. (2007). *Taking Science to School*. Washington, DC: National Academies Press.
[21] Ibid.
[22] National Research Council. (2001). *Adding It Up: Helping Children Learn Mathematics*. Washington, DC: National Academies Press.
[23] W. H. Schmidt, Curtis C. McKnight, and S. Raizen (Eds.). (1997). *A Splintered Vision: An Investigation of U.S. Science and Mathematics Education*. Boston: Kluwer.
[24] Project 2061. (2001). *Atlas of Science Literacy, Volume 1*. Washington, DC: AAAS Press.
[25] National Research Council. (2009). *Learning Science in Informal Environments: People, Places and Pursuits.* Washington, DC: National Academies Press.
[26] K. A. Ericsson. (2006). The Influence of Experience and Deliberate Practice on the Development of Superior Expert Performance. In K. A. Ericsson, N. Charness, P. Feltovich, and R. R. Hoffman (Eds.). *Cambridge Handbook of Expertise and Expert Performance* (pp. 685–706). Cambridge, UK: Cambridge University Press.
[27] K. A. Ericsson and P. Ward. (2007). Capturing the Naturally Occurring Superior Performance of Experts in the Laboratory: Toward a Science of Expert and Exceptional Performance. *Current Directions in Psychological Science,* 16:346–350.
[28] Bogdan Draganski, Christian Gaser, Gerd Kempermann, H. Georg Kuhn, Jürgen Winkler, Christian Büchel, and Arne May. 2006. Temporal and Spatial Dynamics of Brain Structure Changes during Extensive Learning. *The Journal of Neuroscience,* 26(23): 6314– 6317.
[29] National Research Council. (2001). *Adding It Up: Helping Children Learn Mathematics.* Washington, DC: National Academies Press.
[30] National Research Council. (2007). *Taking Science to School: Learning and Teaching Science in Grades K-8.* Washington, DC: National Academies Press.
[31] U.S. Department of Education. (2010). *A Blueprint for Reform: The Reauthorization of the Elementary and Secondary Education Act.* Washington, DC: U.S. Department of Education.

[32] For more information, see http://www.corestandards.org.

[33] Although the formulation of standards for K-12 education in technology-related subject areas has not yet reached the same level of maturity, organizations such as the National Academy of Engineering, the National Science Foundation, the Association for Computing Machinery, the College Board, and the Computer Science Teachers Association have begun to lay the foundation for such standards and for the curricular material on which they might ultimately be based. [For more information, see: National Academy of Engineering. (2009). *Engineering in K-12 Education: Understanding the Status and Improving the Prospects.* Washington, DC: National Academies Press; National Science Foundation. (2009). *Using Computational Thinking to Model a New Course.* Award Number 0938336. Accessible at http://www.nsf.gov/awardsearch/showAward.do?AwardNumber=0938336; Association for Computing Machinery. (2006). *A Model Curriculum for K-12 Computer Science: Final Report of the ACM K-12 Task Force*, Second Edition. New York: Association for Computing Machinery.; College Board. (2010). *AP Computer Science: Principles -- Course Annotations.* New York: College Board.; Computer Science Teachers Association. (2005). *The New Educational Imperative: Improving High School Computer Science Education*. New York: Association for Computing Machinery.]

[34] See Chapter 3 for details.

[35] National Science Board. (2008). *Science and Engineering Indicators: 2008*. Arlington, VA: National Science Foundation.

[36] GAO. (2006). *Science, Technology, Engineering, and Mathematics Trends and the Role of Federal Programs*, statement of Cornelia M. Ashby, Director, Education, Workforce, and Income Security Issues. Accessible at http://www.gao.gov/new.items/d06702t.pdf.

[37] Deloi tte, Oracle, and the Manufacturing Institute. (2009). *People and Profitability: A Time for Change*. Accessible at http://www.deloitte.com/assets

[38] This number is published in the Business and STEM Industry Coalition Charter, March 12, 2010, accessible at http://icw.uschamber.com/sites/default/files/documents/CHARTER%20HANDOUT.pdf. Neither PCAST nor its working group have analyzed whether this is the appropriate number of STEM graduates required for the U.S. economy as this was beyond the scope of this report.

[39] National Association of Colleges and Employers. (2010). *Salary Survey, Winter 2010*. Accessible at http://www.naceweb.org/research/salarysurvey/?referal=research&menuID=71.

[40] Pew Research Center for the People & the Press. (2009). *Public Praises Science; Scientists Fault Public, Media*. Accessible at http://people-press.org/report/?pageid=1546.

[41] Based on numbers from the NCES Department of Education Statistics and National Science Board's *Science and Engineering Indicators 2008*, as cited in the Business-Higher Education Forum (BHEF). (2010). *Increasing the Number of U.S. STEM Graduates: Insights from the STEM Education Modeling Project* (BHEF Working Paper, April 2010, p. 1). Washington, D.C.: Business-Higher Education Forum. Accessible at http://www.stemnetwork.org/research/. Roughly 3.6 million students proceed from the eighth grade. Each year, about 1.5 million bachelor's degrees are conferred. About 240,000 are STEM degrees in fields excluding the social and behavioral sciences (6.7 percent of ninth graders) and about 480,000 are in STEM fields including the social and behavioral sciences. However, the BHEF working paper predicts that of the 4 million ninth-graders in the Nation in 2001, only 4 percent will earn STEM degrees by 2011. Three million students graduated from high school in 2005, 1.9 million attended two or four-year colleges, and only 167,000 are expected to earn STEM degrees within six years of entering college, i.e. by 2011.

[42] National Assessment of Educational Progress. (2010). Results from 2009 in mathematics and most recent science exam. Accessible at http://nces.ed.gov/nationsreportcard/.

[43] Business-Higher Education Forum. (2010). *Increasing the Number of U.S. STEM Graduates: Insights from the STEM Education Modeling Project* (BHEF Working Paper, April 2010, p. 4). Washington, D.C.: Business-Higher Education Forum. Accessible at http://www.stemnetwork.org/research/.

[44] National Center for Education Statistics. (2006). *Digest of Education Statistics: 2005* (NCES 2006-030). Washington, DC: Institute of Education Sciences. Accessible at http://nces.ed.gov/pubsearch/pubsinfo.asp?pubid=2006030.

[45] Higher Education Research Institute Research Brief. (2010). *Degrees of Success*. Accessible at www.heri.ucla.edu/nih/HERIResearchBriefOL2010STEM.pdf.

[46] R. Tai, C. Q. Liu, A. V. Maltese, and X. T. Fan. (2006). Planning for Early Careers in Science. *Science,* 312(5777):1143–1144.

[47] A. V. Maltese and R. H. Tai. (2010). Eyeballs in the Fridge: Sources of Early Interest in Science. *International Journal of Science Education,* 32:669–685.

[48] Although some estimates of this figure may differ, this is based on a rigorous analysis by the PCAST Working Group of documented programs that qualify as K-12 STEM education initiatives.

[49] The American Recovery and Reinvestment Act of 2009 (ARRA) boosted this percentage temporarily to around 11 to 12 percent.

[50] National Center for Education Statistics. (2009). *Digest of Education Statistics*. Washington, DC: U.S. Department of Education. Accessible at http://nces.ed.gov/programs/digest/2009menutables.asp.

[51] Bureau of Labor Statistics. (2010). *Occupational Outlook Handbook, 2010-1 1 Edition*. Washington, DC: U.S. Department of Labor. Accessible at http://www.bls.gov/oco/ocos069.htm.

[52] U.S. Census Bureau. (2010). *Public Education Finances, 2008*. Washington, DC: U.S. Census Bureau.

[53] These percentages do not include one-time funding under the American Recovery and Reinvestment Act of 2009, which as of the publication date of this report had not been fully paid out.

[54] Public Law 103-227. (1994). *GOALS 2000: Educate America Act*. Accessible at http://www2.ed.gov/legislation

[55] Public Law 103-382. (1994). *Improving America's Schools Act of 1994*. Accessible at http://www2.ed.gov/legislation

[56] The America Creating Opportunities to Meaningfully Promote Excellence in Technology, Education, and Science Act (COMPETES) became law in August 2007. While COMPETES authorized many new STEM education programs, few received the appropriations needed to implement them. Reauthorization of this legislation is currently being considered by the Congress.

[57] The ESEA has been renamed several times during reauthorizations. During President George W. Bush's administration, it was called the No Child Left Behind Act. This Act is in the process of being reauthorized, which may lead to various titles being renamed.

[58] Most of the programs mentioned in this section are funded under ESEA. Loan forgiveness is under Federal Student Aid; the Institute for Education Sciences is under the Education Science Reform Act of 2002. Race to the Top and Investing in Innovation (i3) are under ARRA, with plans to include them under the reauthorization of ESEA.

[59] U.S. Department of Education. (2009). *Improving Basic Programs Operated by Local Educational Agencies (Title I, Part A)*. Accessible at http://www2.ed.gov/programs/titleiparta/index.html.

[60] U.S. Department of Education. (2009). *Fiscal Year 2010 Budget Summary, May 7, 2009*. Washington, DC: U.S. Department of Education. Accessible at http://www2.ed.gov/about/overview/budget/budget10/summary/edlite-section3a.html; and U.S Department of Education. (2009). *Fiscal Year 2010 President's Request*. Washington, DC: U.S. Department of Education.

[61] In FY11, the Administration is not requesting separate funding for Educational Technology State Grants but rather is encouraging the infusion of educational technology across programs.

[62] There is documented funding of $156 million for 1,622 grants for science and technology activities (not including tutoring, supplemental education services, or homework help) in the 25 states that report individual (rather than aggregated) activity. Grants are awarded through competition to more than 15 kinds of organizations – local education agencies, community-based organizations, universities, Boys and Girls Clubs, and so on – which report that about 32 percent of their activities target STEM. U.S. Department of Education. (2010). *21st CCLC Funding for STEM Activities: Estimates Based on APR 2009 PPICS Data*. Washington, DC: U.S. Department of Education.

[63] In response to a PCAST request, OMB requested data from Federal agencies on their STEM Education programs. These data were then collected in a central database by the Science and Technology Policy Institute (STPI) and expanded with additional information on the purpose of each of the STEM programs. For this report, PCAST analyzed data in the central database.

[64] Ibid.

[65] The Department is planning to invest even more in strategic STEM programs – at least $606 million if 21st Century Community Learning Centers (CCLC) funding is included. See Table 3-1.

[66] The primary categories included are programs specifically designated as K-12 programs, as well as programs for K-12 teachers (including undergraduate science and mathematics teacher preparation), informal and outreach programs, and research on STEM education.

[67] National Science Board. (2000). *National Science Board: A History in Highlights 1950-2000*. Arlington, VA: National Science Foundation. Accessible at http://www.nsf.gov/nsb/documents/2000/nsb00215/nsb00215.pdf.

[68] Projects funded in the mid-to-late 1980s include Full Option Science System, or FOSS (for elementary grades; an award was made in the late 1990s for a middle school version); Science and Technology for Children, or STC; and ChemComm. FOSS and STC were funded as Triad Projects in the 1980s, and both had middle school versions funded in the 1990s. More recently, Active Chemistry was funded in 2001.

[69] Some notable projects are: Connected Math (originally funded in 1991 and revised in 1999); Everyday Math (materials for grades 3-7 were supported by the NSF); Active Physics (originally funded in 1993 and revised in 2003); Contemporary Mathematics in Context (originally funded in 1992 and revised in 1999); Investigating and Questioning our World through Science and Technology (IQWST; funded in 2004); and Mathematics Instruction using Decision Science and Engineering Tools (funded in 2007).

[70] Both the NSF and the Department of Education fund Mathematics and Science Partnership programs. They differ in that the NSF funding is awarded through a competitive grant process, while the Department of Education funding flows through formula- driven grants to states.

[71] National Science Foundation. (2007). *Merit Review Broader Impacts Criterion: Representative Activities*. Arlington, VA: National Science Foundation. Accessible at http://www.nsf.gov/pubs/gpg/broaderimpacts.pdf.

[72] Figures in this box are based on data received in response to a PCAST request whereby OMB requested data from Federal agencies on their STEM Education programs. These data were then collected in a central database by

the Science and Technology Policy Institute (STPI) and expanded with additional information on the purpose of each of the STEM programs. For this report, PCAST analyzed data in the central database. The other data came directly from NSF.

[73] In response to a PCAST request, OMB requested data from Federal agencies on their STEM Education programs. These data were then collected in a central database by the Science and Technology Policy Institute (STPI) and expanded with additional information on the purpose of each of the STEM programs. For this report, PCAST analyzed data in the central database.

[74] See http://www.nasa.gov/offices/education and http://www.nasa.gov/offices/education

[75] National Research Council. (2008). *NASA's Elementary and Secondary Education Program: Review and Critique.* Washington DC: National Academies Press.

[76] OMB data from HHS/NI H and http://science

[77] In response to a PCAST request, OMB requested data from Federal agencies on their STEM Education programs. These data were then collected in a central database by the Science and Technology Policy Institute (STPI) and expanded with additional information on the purpose of each of the STEM programs. For this report, PCAST analyzed data in the central database.

[78] National Research Council. (2010). *NOAA's Education Program: Review and Critique.* Washington, DC: National Academies Press. Accessible at http://www.nap.edu/catalog/12867.html.

[79] Office of Management and Budget. (2009). *Increased Emphasis on Program Evaluations* (Memorandum M-10-01). Accessible at http://www.whitehouse.gov/sites/default/files/omb/assets. On October 7, 2009, OMB Director Peter Orzag wrote a memorandum to the heads of all executive departments and agencies reporting the launch of a government- wide effort to improve program evaluations. Specifically, "As part of the Fiscal Year 2011 Budget process, OMB will allocate a limited amount of funding for agencies which, on a voluntary basis:

- Show how their Fiscal Year 2011 funding priorities are evidence-based or otherwise subject to rigorous evaluation;
- Assess their own capacity to support evaluation and suggest pathways for strengthening that capacity;
- Propose new evaluations that could improve government programs in the future; and
- Identify impediments to rigorous program evaluation in their statutes or regulations."

[80] Ibid, and testimony of Kathryn Stack to PCAST, October 22, 2009, accessible at www.tvworldwide.com /events/pcast/091022.

[81] This committee had a temporary duration, but its life was extended beyond the March 2009 expiration date.

[82] The America COMPETES Act Reauthorization Act of 2010, 111th Congress, Second Session, H.R. 5116. For full text, see http://frwebgate.access

[83] In the case of technology and engineering, most states have no standards at all. [See: Association for Computing Machinery. (2006). *A Model Curriculum for K-12 Computer Science: Final Report of the ACM K-12 Task Force,* Second Edition. New York: Association for Computing Machinery. See also: National Academy of Engineering. (2009). *Engineering in K-12 Education: Understanding the Status and Improving the Prospects.* Washington, DC: National Academies Press.]

[84] National Mathematics Advisory Panel. (2008). *Foundations for Success: The Final Report of the National Mathematics Advisory Panel.* Washington, DC: Department of Education. Accessible at http://www2.ed .gov/about/bdscomm/list/mathpanel/report/final-report.pdf.

[85] The National Mathematics Advisory Panel (2008) concluded that while the NAEP framework is sufficient to support the main conclusions about national and states' progress in mathematics since 1990, it does not provide enough guidance to test developers or measurement precision for low-achieving students. Furthermore, the NAEP mathematics assessment could benefit from better quality of and more evidence for test item design, including a move in the direction of computerized adaptive testing. National Mathematics Advisory Panel. (2008). *Foundations for Success: The Final Report of the National Mathematics Advisory Panel.* Washington, DC: U.S. Department of Education.

[86] Project 2061 was launched in 1985, the last time Haley's Comet was visible from Earth. The next time it will reappear is in 2061, hence the project's name.

[87] National Council of Teachers of Mathematics. (2006). *Curriculum Focal Points for Prekindergarten through Grade 8 Mathematics: A Quest for Coherence.* Reston, VA: NCTM. Accessible at http://www.nctm.org /catalog/product.aspx?ID=13089.

[88] Commission on Mathematics and Science Education. (2009). *The Opportunity Equation. Transforming Mathematics and Science Education for Citizenship and the Global Economy.* Carnegie Corporation of New York and the Institute for Advanced Study.

[89] National Governors Association, Council of Chief State School Officers, and Achieve, Inc. (2008). *Benchmarking for Success: Ensuring U.S. Students Receive A World-Class Education.* Washington, DC: National Governors Association.

[90] The NRC's science education framework committee's work is funded by the Carnegie Corporation of New York.

[91] National Academy of Engineering. (2009). *Engineering in K-12 Education: Understanding the Status and Improving the Prospects.* Washington, DC: National Academies Press.

[92] Association for Computing Machinery. (2006). *A Model Curriculum for K-12 Computer Science: Final Report of the ACM K-12 Task Force Second Edition.* New York, NY: Association for Computing Machinery.

[93] NAEP is developing pilot technology literacy assessments. A framework for this work was approved by the National Assessment Governing Board in May 2010. The assessment will be administered in 2014. For more information, see http://www.edgateway.net/cs/naepsci/print/docs/470.

[94] On July 12, 2010, the National Research Council released for public comment a draft of a framework for science education standards. The draft includes a three-dimensional framework. The first dimension describes disciplinary core content, including core ideas in the life science, earth sciences, physical sciences, and engineering and technology. The second dimension includes cross-cutting elements with applicability across science disciplines. The third dimension describes science and engineering practices.

[95] Association for Computing Machinery. (2006). *A Model Curriculum for K-12 Computer Science: Final Report of the ACM K-12 Task Force Second Edition*. New York, NY: Association for Computing Machinery.

[96] Ibid.

[97] Computer Science Teachers Association. (2005). *The New Educational Imperative: Improving High School Computer Science Education.* New York: Association for Computing Machinery.

[98] For further information on the Advanced Placement CS: Principles course, see http://csprinciples.org.

[99] National Academy of Engineering. (2009). *Engineering in K-12 Education: Understanding the Status and Improving the Prospects.* Washington, DC: National Academies Press.

[100] Paul Bambrick-Santoyo. (2010). *Driven by Data: A Practical Guide to Improve Instruction.* San Francisco: Jossey-Bass.

[101] B. Topol, J. Olson, and E. Roeber. (2010). *The Cost of New Higher Quality Assessments: A Comprehensive Analysis of the Potential Costs for Future State Assessments.* Stanford, CA: Stanford University, Stanford Center for Opportunity Policy in Education. In a four-year cycle, Year 0 includes development costs for the assessment program; Years 1-3 are the operation of a conventional comprehensive assessment program.

[102] Ibid.

[103] Topol et al. define a "moderately sized state" as having about 125,000 students per grade tested in grades 3-8, and 10. Students are tested in mathematics, reading, and writing using a pencil and paper summative assessment. Ibid, p. 12.

[104] The costs to develop high-quality science assessments are unknown, but given the need to reflect more aspects of learning and a broad range of topics, it is possible they will cost much more to develop and administer than assessments for mathematics and reading.

[105] With multistate consortia, representing about 30 states or more, the high-quality assessment cost is likely to be closer to $39 per student for mathematics, reading, and writing combined.

[106] If high-quality assessments for mathematics, reading, and writing were delivered online, Topol et al. estimate that the cost could drop by $2.25 and $3.50 per student, depending on whether multistate consortia adopt them or single states do. Nevertheless, upfront development costs could be much higher than for traditional test formats.

[107] Excerpt from page 25, Office of Educational Technology, U.S. Department of Education. (2010). *Transforming American Education: Learning Powered by Technology.* Washington, DC: U.S. Department of Education. Accessible at http://www.ed.gov/sites/default/files/NETP-2010-final-report.pdf.

[108] B. Topol, J. Olson, and E. Roeber. (2010). *The Cost of New Higher Quality Assessments: A Comprehensive Analysis of the Potential Costs for Future State Assessments.* Stanford, CA: Stanford University, Stanford Center for Opportunity Policy in Education.

[109] Evaluation studies of the NSF Local Systemic Change initiatives found that professional development of about 130 hours duration, aligned to the specific curriculum to be taught in schools, cost $3,500 to $4,800 per teacher, spread out over several years of teaching. See Eric R. Banilower, Sally E. Boyd, Joan D. Pasley, and Iris R. Weiss. (2006). *Lessons from a Decade of Mathematics and Science Reform: A Capstone Report for the Local Systemic Change through Teacher Enhancement Initiative.* Chapel Hill, NC: Horizon Research.

[110] For mathematics, at least $1 billion annually will be required to support high-quality assessments: approximately 3.8 million students per grade tested time seven grades time $39 per student (presuming assessments are developed by 30-state consortia), which equals $1.04 billion per year.

[111] Jill Leovy. (2010). Honoring a Legendary Teacher and His Legacy. *The Los Angeles Times*, April 17, 2010. Accessible at http://articles.latimes.com/2010/apr/17/local/la-me-escalante17-2010apr17.

[112] National Research Council. (2010). *Preparing Teachers: Building Evidence for Sound Policy*. Washington, DC: National Academies Press.

[113] D. Aaronson, L. Barrow, W. L. and Sander. (2003). *Teachers and Student Achievement in the Chicago Public High Schools.* Chicago, IL: Research Department, Federal Reserve Bank of Chicago.

[114] W. L. Sander and J. C. Rivers. (1996). *Cumulative and Residual Effects of Teachers on Future Student Academic Achievement.* Knoxville, TN: University of Tennessee Value-Added Research and Assessment Center. Accessible at http://www.mccsc.edu/~curriculum

[115] National Mathematics Advisory Panel. (2008). *Foundations for Success: The Final Report of the National Mathematics Advisory Panel.* Washington, DC: U.S. Department of Education.

[116] S. G. Rivkin, E. A. Hanushek, and J. F. Kain. (2005). Teachers, Schools, and Academic Achievement. *Econometrica, 73(2)*:417– 458. Accessible at http://www.econ.ucsb.edu/~jon/Econ230C/HanushekRivkin.pdf.

[117] National Mathematics Advisory Panel. (2008). *Foundations for Success: The Final Report of the National Mathematics Advisory Panel.* Washington, DC: U.S. Department of Education. More recent studies suggest an even larger effect size.

[118] Ibid.

[120] Heather C Hill; Brian Rowan; Deborah L. Ball. (2005). Effects of Teachers' Mathematical Knowledge for Teaching on Student Achievement. *American Educational Research Journal,* 42:371.

[121] National Research Council. (2010). *Preparing Teachers: Building Evidence for Sound Policy.* Washington, DC: National Academies Press.

[122] Ibid, Chapter 7.

[123] National Research Council. (1999). *How People Learn: Bridging Research and Practice*, and *How People Learn: Brain, Mind, Experience and School: Expanded Edition.* Washington, DC: National Academies Press.

[124] A. V. Maltese and R. H. Tai, R.H. (2010). Eyeballs in the Fridge: Sources of Early Interest in Science. *International Journal of Science Education,* 32: 669-685.

[125] Two reports of note are *Adding it Up*, National Research Council, 2001 and *Taking Science to School*, National Research Council 2007.

[126] Richard Ingersoll. (n. d.). Unpublished analyses of the 2007-08 Schools and Staffing Survey data that spans K-12.

[127] R. Ingersoll and D. Perda, D. (2010). Is The Supply of Mathematics and Science Teachers Sufficient? *American Educational Research Journal,* 47(3): 563-594. Accessible at http://aer.sagepub.com/content/early/2010/05/13/0002831210370711.

[128] This is not a STEM specific figure, but Richard Ingersoll of the University of Pennsylvania says this is probably a fair assumption given that turnover rates are comparable to other teaching subjects.

[129] R. Ingersoll and D. Perda, D. (2010). Is The Supply of Mathematics and Science Teachers Sufficient? *American Educational Research Journal,* 47(3): 563-594. Accessible at http://aer.sagepub.com/content/early/2010/05/13/0002831210370711.

[130] Ingersoll and Perda. (2010). Pg. 124.

[131] Ingersoll and Perda. (2010). P. 146 and 151.

[132] R. Ingersoll. (2008). *Core problems: Out-of-field teaching persists in key academic courses and high-poverty schools.* Washington, DC: Education Trust.

[133] National Research Council. (2010). *Preparing Teachers: Building Evidence for Sound Policy.* Washington, DC: National Academies Press.

[134] Ibid, Chapter 3.

[135] National Academy of Education. (2009). *Education Policy White Paper: Science and Mathematics Education.* Washington DC: National Academy of Education. Accessible at www.naeducation.org/NAEdWhitePapersProject.html.

[136] Ibid.

[137] Nevertheless, in particular schools and school districts, middle school teacher preparation is comparable to that of elementary teachers. See Center for Research in Mathematics and Science Education. (2010). *Breaking the Cycle: An International Comparison of U.S. Mathematics Teacher Preparation, Initial Findings for the Teacher Education and Development Study in Mathematics.* Lansing, MI: Michigan State University. Accessible at http://www.educ.msu.edu/content/default.asp?contentID=710.

[138] Unpublished analyses of the 2007-08 Schools and Staffing Survey data by Richard Ingersoll, University of Pennsylvania. In the 2007-08 school year, 476,922 public school teachers in K-12 had a main teaching assignment in math or science. 51,000 were employed in primary schools, and 426,000 in middle, secondary or combine schools. Of the middle and high school teachers, 256,000 (60 percent) had a full degree in mathematics, a scientific discipline, mathematics education, or science education. 146,000 of those teachers had only a degree from an education department or school. 57,500 of the middle and high school teachers had a degree from a science or mathematics department or school, and 52,400 had a degree from both an education department or school and a science or mathematics department or school.

[139] National Mathematics Advisory Panel. (2008). *Foundations for Success: The Final Report of the National Mathematics Advisory Panel.* U.S. Department of Education: Washington, DC. On p. 40, the report states: "Despite widespread beliefs about the qualities that make teacher education effective, the Panel did not find strong evidence for the impact of any specific form of, or approach to, teacher education on either teachers' knowledge or students' learning. Even for the few studies that did produce significant effects, little detail was provided about the features of the training that might account for the impact of the program."

[140] R. Ingersoll and D. Perda, D. (2010). Is The Supply of Mathematics and Science Teachers Sufficient? *American Educational Research Journal,* 47(3): 563-594. Accessible at http://aer.sagepub.com/content/early/2010/05/13/0002831210370711.

[141] National Association of Colleges and Employers. (2010). *Salary Survey, Spring 2010.* Accessible at http://www.naceweb.org/research/salarysurvey/?referal=research&menuID=71.
[142] Ibid.
[143] National Science Board. (2008). *Science and Engineering Indicators: 2008.* Arlington, VA: National Science Foundation.
[144] Organisation for Economic Co-operation and Development. (2009). *Education at a Glance 2009: OECD Indicators.* Washington, DC: OECD. Accessible at_http://www.oecd.org/edu/eag2009, see table under Indicator D4.
[145] Ibid, see Indicator D3.
[146] This figure is based on most recent data from the Robert Noyce Scholarship program at the NSF, which notes that in 2008-09, 871 prospective STEM teachers were supported with an average scholarship of $7,924, and from the Annenberg Fellowship Program, which as of 2009 awarded 80 prospective STEM teachers per year with scholarships of $30,000 to complete a year-long master's program in teaching.
[147] In its FY11 budget request, the administration proposes a $950 million Teacher and Leader Innovation Fund, which would make competitive awards to states and local education agencies willing to implement bold approaches to improving the effectiveness of the education workforce in high-need schools by creating the conditions needed to identify, reward, retain, and advance effective teachers, principals, and school leadership teams in those schools, and enabling schools to build the strongest teams possible. (Note: language is from FY11 budget request.)
[148] *The STEM Master Teacher Corps would differ from existing teacher recognition programs because it would recognize a larger percentage of teachers, would be an interactive community, and would provide a larger monetary reward.* Inspired math and science teachers have been recognized through the Presidential Awards for Excellence in Mathematics and Science Teaching (PAEMST) administered by the NSF, the Robert Noyce Scholarship program administered by the NSF, and the Albert Einstein Distinguished Educator fellowship program administered by the Department of Energy. In addition, the nonprofit organization National Board for Professional Teaching Standards (NBPTS) offers an advanced teaching credential (including in mathematics and in science for students ages 11–18+). This credential is voluntary and requires that teachers develop a rigorous portfolio and participate in an assessment, which costs $2,500. Through 2009, more than 6,000 credentials have been awarded in early adolescent (ages 11–14) or adolescent (ages 15–18+) mathematics, and nearly the same number in science at the same levels. The amount of local and state support for the assessment fees and for annual stipends to those who receive the credential varies widely. In some states, teachers' fees are paid and they receive annual stipends of up to $5,000, which is often tied to working in high-needs schools. In other states and districts, teachers receive no recognition or reward. In both of these programs, teachers are recognized for their professionalism and accomplishments. A STEM master teacher corps would envision a more enhanced, sustained role for teachers, for example, through a national network of teachers that can serve as mentors, leaders in their schools, and liaisons to public officials, or as excellent teachers who would be connected with their peers to share best practices and materials, or to provide a voice for their profession in educational policy and to be effective advocates for STEM in their schools, school districts, and communities.
[149] Office of Educational Technology, U.S. Department of Education. (2010). *Transforming American Education: Learning Powered by Technology*. Washington, DC: U.S. Department of Education. Accessible at http://www.ed.gov/sites/default/files/NETP-2010- final-report. pdf.
[150] Ibid, pp. 25-38.
[151] Ibid.
[152] National Science Foundation Task Force on Cyberlearning. (2008). *Fostering Learning in the Networked World: The Cyberlearning Opportunity and Challenge: A 21st-Century Agenda for the National Science Foundation.* Arlington, VA: National Science Foundation.
[153] Federal Communications Commission. (2010). *Connecting America: The National Broadband Plan*. Washington, DC: Federal Communications Commission.
[154] Ibid, pp. 127–136.
[155] Robert Tinker (personal communication, 2010).
[156] i EARN was founded by a retired businessman. The earliest work on Logo at BBN was funded by corporate funds. Mastering Physics was developed by an MIT professor.
[157] Department of Education, Institute of Education Sciences. (2007). *What Works Clearinghouse: Cognitive Tutor Algebra I.* Washington, DC: Department of Education. Accessible at_http://ies.ed.gov/ncee/wwc/pdf/WWCCognitiveTutor052907.pdf.
[158] C. Christensen, C. W. Johnson, and M. B. Horn. (2008). *Disrupting Class: How Disruptive Innovation Will Change the Way the World Learns*. New York: McGraw-Hill.
[159] See, for instance, http://learningcenter.nsta.org/products/onlinecourses/Default.aspx#nsta and http://scienceonline.terc.edu.

[160] Office of Educational Technology, U.S. Department of Education. (2010). *Transforming American Education: Learning Powered by Technology*. Washington, DC: U.S. Department of Education. Accessible at http://www.ed.gov/sites/default/files/NETP-2010- final-report. pdf.

[161] Merrilea J. Mayo. (2009). *Bringing Game-Based Learning to Scale: The Business Challenges of Serious Games*. Paper presented at Workshop on Learning Science: Computer Games, Simulations, and Education. Washington, DC: October 6–7, 2009. Accessible at_http://www7.nationalacademies.org/bose/MayoGamingCommissionedPaper.pdf.

[162] D. Clark, B. Nelson, P. Sengupta, and C. D'Angelo. (2009). *Rethinking Science 1 Learning Through Digital Games and 2 Simulations: Genres, Examples, and Evidence*. Paper presented at Workshop on Learning Science: Computer Games, Simulations, and Education, Washington, DC, October 6-7, 2009. Accessible at http://www7.nationalacademies.org/bose/ClarkGamingCommissionedPaper.pdf.

[163] For an example, see_http://www7.nationalacademies.org/bose/FletcherGamingPresentation.pdf.

[164] This list only illustrates some of the many categories and examples of ways technology can enhance STEM education. For more information, see Andrew A. Zucker. (2008). *Transforming Schools with Technology: How Smart Use of Digital Tools Helps Achieve Six Key Education Goals*. Cambridge, MA: Harvard; and L. Johnson, A. Levine, R. Smith, and S. Stone. (2010). *The 2010 Horizon Report*. Austin, Texas: The New Media Consortium. For compilations of exemplary programs, see http://computedgazette.com/index.html; http://www.sciencemag.org/special/spore and http://nsdl.org.

[165] See http://www.kaputcenter.umassd.edu/projects/simcalc.

[166] See http://www.pasco.com/featured-products/airlink-si/index.cfm.

[167] "PINPoint" features a range of on-board sensors, a GPS receiver, and a connector for external sensors. F. Martin, S. Kuhn, M. Scribner-MacLean, C. Corcoran, J. Dalphond, J. Fertitta, M. McGuinness, S. Christy, and I. Rudnicki. (2010). *iSENSE: A Web Environment and Hardware Platform for Data Sharing and Citizen Science*. Presented at the AAAI 2010 Spring Symposium, Educational Robotics and Beyond: Design and Evaluation, Stanford, CA, March 2010.

[168] For more information, see http://mw.concord.org.

[169] For more information, see http://el.media

[170] For more information, see http://ccl.northwestern.edu/netlogo/models

[171] For more information, see http://iearn.org.

[172] One of the most carefully studied projects of this sort is Nancy Songer's research on Kids and Global Scientists and BioKids. For more information, see http://www.biokids.umich.edu and http://drdc.uchicago .edu/community

[173] For more information, see http://wise4.telscenter.org.

[174] Federal Communications Commission. (2010). *Connecting America: The National Broadband Plan*. Washington, DC: Federal Communications Commission.

[175] Alison Gopnik, Andrew Meltzoff, and Patricia Kuhl. (1999). *Scientist in the Crib: What Early Learning Tells Us About the Mind*. New York: HarperCollins.

[176] A. V. Maltese and Tai, R.H. (2010). Eyeballs in the Fridge: Sources of Early Interest in Science. *International Journal of Science Education*, 32:669-685.

[177] István Hargittai. (2002). *The Road to Stockholm: Nobel Prizes, Science, and Scientists*. New York: Oxford University Press.

[178] Donald G. Rhea and Katherine M. Nielsen. (2010). A Volunteer Army for Science. *Science*, 329:257.

[179] For more information about the range of programs offered by these organizations, see the information compiled by the Coalition for Science After School (SAS), which represents 1,400 organizations and over 5,000 after after-school programs with a STEM focus in all 50 states. Its website can be accessed at http://afterschoolscience.org.

[180] D. L. Vandell, E. R. Reisner, and K. M. Pierce. (2007). *Outcomes Linked to High-Quality Afterschool Programs: Longitudinal findings from the Study of Promising Afterschool Programs*. Irvine, California: University of California.

[181] P. Bell, B. Lewenstein, A. W. Shouse, and M. A. Feder (Eds.). (2009). Diversity and equity. Chapter 7. In *Learning Science in Informal Environments: People, Places, and Pursuits*. Washington, DC: National Academies Press. See also http://www.girlscouts.org/program/programopportunities/science_http://hestec.utpa edu, http://mesa.ucop.edu.

[182] P. A. Lauer, M. Akiba, S. B. Wilkerson, H. S. Apthorp, D. Snow, and M. Martin-Glenn. (2006). Out-of-School-Time Programs: A Meta-Analysis of Effects for At-Risk Students. *Review of Educational Research*, 76:275–313.

[183] Nicholas Colangelo and Gary A. Davis (Eds.). (2005). *Handbook on Gifted Education*. Boston: Allyn and Bacon.

[184] Alexander W. Astin. (1971). The pre-college student science training program of the National Science Foundation: An empirical analysis. *The Journal of Experimental Education*, 39:1-12.

[185] Learning Point Associates. (2007). *21st Century Community Learning Centers (21st CCLC) Analytic Support for Evaluation and Program Monitoring: An Overview of the 21st CCLC Performance Data: 2005–06*. Chicago, IL: Learning Point Associates.

[186] Ibid, Chapter 3.

[187] INPSIRE could, for example, serve as an acronym for: Individualized STEM programs to Interest, Re-engage, and Educate.

[188] National Science Board. (2010). *Science and Engineering Indicators: 2010*. Arlington, VA: National Science Foundation. Accessible at http://www.nsf.gov/statistics

[189] C. Kirabo Jackson. (2010). *A Stitch in Time: The Effects of a Novel Incentive-Based High-School Intervention on College Outcomes* (NBER Working Paper No. 15722). Cambridge, MA: National Bureau of Economic Research. Accessible at http://www.nber.org/papers/w15722.pdf.

[190] See http://host

[191] Michael S. Holstead, Terry E. Spradlin, Margaret E. McGillivray, and Nathan Burroughs. (2010). The Impact of Advanced Placement Incentive Programs. *Center for Evaluation and Education Policy Education Policy Brief*, 8(1):1-12.

[192] See C. Dougherty, L. Mellor, S. Jian. (2006). *The Relationship Between Advanced Placement and College Graduation*. National Center for Educational Accountability: 2005 NCEA Study Series, Report 1, Web, 4 May 2010. Accessible at http://www.nc4ea.org/files/relationshipbetweenapandcollegegraduation02-09-06.pdf. (Noting that schools should use AP exams, "not as a special set of courses for their already well-prepared students, but as a comprehensive program to prepare large numbers of students, starting in the early grades and including disadvantaged students, to be able to do college-level work before they leave high school"); see also K. Klopfenstein. (2004). The Advanced Placement Expansion of the 1990s: How Did Traditionally Underserved Students Fare. *Education Policy Analysis Archives*, 12(68). Accessible at_http://epaa.asu.edu/epaa/v12n68/. (Analyzing the need to increase access to AP classes for traditionally underrepresented students).

[193] We note that not only seniors take these AP exams, but nonetheless it provides useful context to observe the relatively small percentage of graduating seniors that the students who take and pass AP exams represents.

[194] National Consortium for Specialized Secondary Schools for Mathematics, Science and Technology (personal communication, 2010).

[195] A recent study of 22 KIPP schools showed their potential for success in mathematics education: 18 of the schools raised student achievement in mathematics significantly relative to their district counterparts, and, in a three-year-period, half of the schools raised students' mathematics achievement on state assessments enough, on average, to exceed the national norms for student growth by the equivalent of 1.2 years of additional instruction. Tuttle, Christina C, et al. (2010). *Student Characteristics and Achievement in 22 KIPP Middle Schools: Final Report*. Washington, DC: Mathematica Policy Research, Inc. Accessible at www.mathematica-mpr.com/publications/PDFs/education

[196] Research pointing to the synergy between science education and literacy was highlighted in the April 23, 2010, edition of the journal *Science*.

[197] R. F. Subotnik, H. Robert, R. Rickoff, and J. Almarode. (2010). Specialized Public High Schools of Science, Mathematics and Technology and the STEM Pipeline: What Do We Know Now and What Will We Know in 5 Years? *Roeper Review*, 32:7–16.

[198] A. J. Levy, M. M. Pasquale, and L. Marco. (2008). Models of Providing Instruction in the Elementary Grades: A Research Agenda to Inform Decision Makers. *Science Educator*, 17(2):1–18.

In: U.S. Science, Technology, Engineering and Math (STEM)... ISBN: 978-1-61209-792-3
Editor: Jason C. Rollins

Chapter 2

PREPARING THE NEXT GENERATION OF STEM INNOVATORS: IDENTIFYING AND DEVELOPING OUR NATION'S HUMAN CAPITAL

National Science Board

MEMORANDUM FROM THE CHAIRMAN OF THE NATIONAL SCIENCE BOARD

SUBJECT: *Preparing the Next Generation of STEM Innovators: Identifying and Developing our Nation's Human Capital*

Scientific and technological innovation continues to play an essential role in catalyzing the creation of new industries, spawning job growth, and improving the quality of life in the United States and throughout the world. Innovation relies, in part, on individuals possessing the knowledge, skills, creativity, and foresight to forge new paths. The National Science Board (Board) is pleased to present its recommendations on how to support the identification and development of talented young men and women who have the potential to become our Country's next generation of science, technology, engineering, and mathematics (STEM) innovators.

The Board embarked on this detailed study for two mutually reinforcing reasons:

1. The long-term prosperity of our Nation will increasingly rely on talented and motivated individuals who will comprise the vanguard of scientific and technological innovation; and
2. Every student in America deserves the opportunity to achieve his or her full potential.

This report contains a series of policy actions, a research agenda, and three key recommendations detailing how our Nation might foster the identification and development of future STEM innovators. This report draws on the findings from an expert panel discussion held at the National Science Foundation (NSF) on August 23-25, 2009, and a 2-year examination of the issue by the Board with the support of expert staff from the NSF Directorate for Education and Human Resources and the U.S. Department of Education.

The Board firmly believes that the recommendations set forth in this report will help ensure a legacy of continued prosperity and will engender a renewed aspiration towards equity *and* excellence in U.S. STEM education.

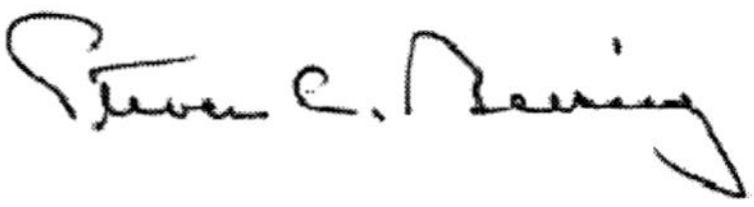

Steven C. Beering
Chairman

ACKNOWLEDGMENTS

The National Science Board (Board) appreciates the numerous individuals who contributed to the work of the Board's Committee on Education and Human Resources and the *ad hoc* Task Group on STEM Innovators. A list of distinguished panelists and discussants who participated in the August 2009 expert panel discussion and provided significant input into the development of the report is provided in Appendix II.

We are particularly indebted to Dr. Cora B. Marrett, Acting Deputy Director of the National Science Foundation (NSF), and the following NSF staff members in the Directorate for Education and Human Resources for assistance in planning the expert panel discussion and/or providing input during the development of the report: Drs. Myles Boylan, Alphonse DeSena, Janice Earle, Joan Ferrini-Mundy, Ping Ge, James E. Hamos, Karen Oates, Ginger Holmes Rowell, and Larry E. Suter. The Board would like to acknowledge the efforts of Ms. Patricia Johnson, U.S. Department of Education, for assistance in planning the expert panel discussion, serving as a panelist, and reviewing several drafts of the report. The Board would like to thank several external experts who provided a critical reading of the report draft: Drs. Linda E. Brody, Carolyn M. Callahan, Nancy Green, Sidney Moon, Paula Olszewski-Kubilius, Sally M. Reis, Nancy M. Robinson, and Mark Saul.

The National Science Board Office (NSBO) provided essential support to the work of the Task Group on STEM Innovators. Especially deserving of recognition are: Dr. Matthew Wilson, AAAS Science and Technology Policy Fellow and NSBO staff lead for the STEM Innovators project, for his thoughtful and diligent work throughout the duration of this initiative; Mses. Jennie Moehlmann and Jean Pomeroy, for policy guidance and critical review of numerous drafts of the report; Ms. Jennifer Richards, for editorial support throughout the drafting process; Mses. Betty Wong, Pamela McKinley, and Kyscha Slater-Williams, for providing administrative support for the expert panel discussion; and Mses. Ann Ferrante and Kelly DuBose for editorial and publishing assistance. Lastly, Dr. Craig R. Robinson, Acting Executive Officer of the Board and Board Office Director, provided essential guidance and support throughout the duration of the project.

PROCESS FOR PRODUCING THE REPORT

The National Science Board (Board) has long been concerned with the state of science, technology, engineering, and mathematics (STEM) education in the United States. In October 2007, the Board asserted in its *National Action Plan for Addressing the Critical Needs of the U.S. Science, Technology, Engineering and Mathematics Education System* (STEM Action Plan, NSB-07-114) that the Nation must enhance its "ability to produce a numerate and scientifically and technologically literate society and to increase and improve the STEM education workforce." In that report and others (e.g., *The Science and Engineering Workforce: Realizing America's Potential*, NSB-03-69), the Board acknowledged that the United States has become increasingly dependent on importing STEM talent from other countries, rather than expanding the STEM pipeline from our own domestic talent pool.

In this report, the Board addresses the educational needs of our Nation's most talented and motivated students, who have the potential to become high-achieving members of the U.S. STEM workforce, or STEM innovators. **STEM "innovators" are defined as those individuals who have developed the expertise to become leading STEM professionals and perhaps the creators of significant breakthroughs or advances in scientific and technological understanding.** To this end, this report addresses talent identification and development of children and young adults, and provides recommendations that should ultimately enhance the innovation capacity of our Nation.

To produce this report, the Board charged the Committee on Education and Human Resources to form an *ad hoc* Task Group on STEM Innovators in August 2008 (see Appendix I). The *ad hoc* Task Group was directed to identify strategies for increasing the number of future STEM innovators and synthesize recommendations for how the National Science Foundation (NSF), and possibly other Federal entities, might engage in fostering the development of these individuals. This report and the recommendations set forth herein are based on the findings from an expert panel discussion held on August 23-25, 2009 (see Appendix III), and a 2-year examination of the issue by the *ad hoc* Task Group with the support of experts from the NSF Directorate for Education and Human Resources and the U.S. Department of Education.

EXECUTIVE SUMMARY

On November 17, 1944, in the midst of World War II, President Franklin Delano Roosevelt wrote a letter to Vannevar Bush, head of the U.S. Office for Scientific Research and Development. In that letter, President Roosevelt posed the question:

> *Can an effective program be proposed for discovering and developing scientific talent in American youth so that the continuing future of scientific research in this country may be assured on a level comparable to what has been done during the war?*[1]

In *Science–The Endless Frontier*, Vannevar Bush offered his answer to this question. In his response, Bush called for the renewal of our scientific talent through the U.S. education system. He wrote:

The responsibility for the creation of new scientific knowledge rests on that small body of men and women who understand the fundamental laws of nature and are skilled in the techniques of scientific research. While there will always be the rare individual who will rise to the top without benefit of formal education and training, he is the exception and even he might make a more notable contribution if he had the benefit of the best education we have to offer.[2]

A little more than a decade later, mobilized by the Soviet's successful launch of Sputnik, the United States embarked on a collective, coordinated, and sustained effort to recruit and educate the "best and brightest" who subsequently would form a new generation of leaders and innovators in science and engineering. This effort resulted in unprecedented scientific and technological progress, leading to the creation of new enterprises, new jobs, and the betterment of the national standard of living. At the root of this progress was a substantial investment in research and development, along with a nationwide focus on excellence in science, technology, engineering, and mathematics (STEM) education and talent development. Regrettably, by the 1970s, this national sense of urgency had diminished, and complacency soon supplanted the ideal of excellence in education. Today, faced with growing international competition, the cost of inaction continues to grow at an intensifying pace.

The National Science Board (Board) firmly believes that to ensure the long-term prosperity of our Nation, we must renew our collective commitment to excellence in education and the development of scientific talent. Currently, far too many of America's best and brightest young men and women go unrecognized and underdeveloped, and, thus, fail to reach their full potential. This represents a loss for both the individual *and* society. The Nation needs "STEM innovators"—those individuals who have developed the expertise to become leading STEM professionals and perhaps the creators of significant breakthroughs or advances in scientific and technological understanding. A key component of innovation is the development of new products, services, and processes essential to the Nation's international leadership. Just as in generations past, there are talented students from every demographic and from every part of our Country who with hard work and with the proper opportunities will form the next generation of STEM innovators. The vital importance of innovation to the U.S. economy led the Board to embark on a 2-year exploration of this issue.

Our analyses of research and demographic data, as well as our consultation with a wide range of experts, practitioners, policy-makers, and STEM innovators, led us to identify three major areas where focused attention is essential. First, while there are some examples of high-impact educational policies and practices that are effective in enabling tomorrow's potential STEM innovators to thrive, many more are needed. Second, a commitment to equity and diversity, and analyses of demographic trends, lead to the conclusion that new, ambitious efforts to cast a wide net in seeking and inspiring tomorrow's STEM leaders are critical. Finally, it is clear that when the learning environment is infused with high expectations and a commitment to excellence, the potential for future innovators to flourish is great.

To identify and develop the next generation of STEM innovators, the Board makes three *keystone recommendations*. Each recommendation contains several ***policy actions*** for the National Science Foundation (NSF), other Federal entities, and the Nation. Additionally, for each keystone recommendation, the Board proposes a ***research agenda*** for NSF that will ensure the policy actions are supported by the best available research. The keystone recommendations and a summary of the policy actions are listed below. The findings and research agenda can be found in the main body of the report (pp. 15-25).

Keystone Recommendations

I. ***Provide opportunities for excellence.*** We cannot assume that our Nation's most talented students will succeed on their own. Instead, we must offer coordinated, proactive, sustained formal and informal interventions to develop their abilities. Students should learn at a pace, depth, and breadth commensurate with their talents and interests and in a fashion that elicits engagement, intellectual curiosity, and creative problem solving—essential skills for future innovation.

To achieve this goal, the Board proposes the following policy actions:

A. Encourage states and/or local education agencies to adopt consistent and appropriate policies on differentiated instruction, curriculum acceleration, and enrichment, and to recognize the achievement levels of students moving or transitioning to different schools.

B. Increase access to and quality of college-level, dual enrollment, and other accelerated coursework, as well as high-quality enrichment programs.

C. Support rigorous, research-based STEM preparation for teachers, particularly general education teachers, who have the most contact with potential STEM innovators at young ages.

D. Provide Federal support to formal and informal programs that have a proven record of accomplishment in stimulating potential STEM innovators.

E. Leverage NSF's *Broader Impacts Criterion* to encourage large-scale, sustained partnerships among higher education institutions, museums, industry, content developers and providers, research laboratories and centers, and elementary, middle, and high schools to deploy the Nation's science assets in ways that engage tomorrow's STEM innovators.

F. Create NSF programs that offer portable, merit-based scholarships for talented middle and high school students to participate in challenging enrichment activities.

G. Increase the technological capabilities and network infrastructure in rural and low-income areas, and expand cyber-learning opportunities.

H. Create a national database of formal and informal education opportunities for highly talented students, and publicize and promote such opportunities nationally to parents, education professionals, and content and resource providers.

II. ***Cast a wide net*** to identify *all* types of talents and to nurture potential in *all* demographics of students. To this end, we must develop and implement appropriate talent assessments at multiple grade levels and prepare educators to recognize potential, particularly among those individuals who have not been given adequate opportunities to transform their potential into academic achievement.

To achieve this goal, the Board proposes the following policy actions:

A. Improve pervasiveness and vertical coherence of existing talent assessment systems.

B. Expand existing talent assessment tests and identification strategies to the three primary abilities (quantitative/mathematical, verbal, and spatial) so that spatial talent is not neglected.
C. Increase access to appropriate above-level tests and student identification mechanisms, especially in economically disadvantaged urban and rural areas.
D. Encourage pre-service education and professional development for education professionals (including teachers, principals, and counselors) in the area of talent identification and development.
E. Encourage pediatricians and early childhood educators, especially *Head Start* teachers, to become knowledgeable about early signs of talent and the need for its nurturance.

III. ***Foster a supportive ecosystem*** that nurtures and celebrates excellence and innovative thinking. Parents/guardians, education professionals, peers, and students themselves must work together to create a culture that expects excellence, encourages creativity, and rewards the successes of all students regardless of their race/ethnicity, gender, socioeconomic status, or geographical locale.

To achieve this goal, the Board proposes the following policy actions:

A. Create a national campaign aimed at increasing the appreciation of academic excellence and transforming stereotypes towards potential STEM innovators.
B. Encourage the creation of positive school environments that foster excellence by providing professional development opportunities for teachers, principals, counselors, and other key school staff.
C. Support the expansion of computing and communications infrastructure in elementary, middle, and high schools to foster peer-to-peer connections and collaborations, and direct connections with the scientific research community.
D. Hold schools, and perhaps districts and states, accountable for the performance of the very top students at each grade.
E. Have NSF, in partnership with the Institute of Education Sciences, hold a high-level conference to bring together researchers in the learning sciences, other scientists, education school administrators, current teachers and principals, and teacher professional associations to discuss teacher preparation and pedagogical best practices aimed at fostering innovative thinking in children and in young adults.

The United States is faced with a clear and profound choice between action and complacency. The Board firmly believes that a coherent, proactive, and sustained effort to identify and develop our Nation's STEM innovators will help drive future economic prosperity and improve the quality of life for all. Likewise, providing opportunities for all young men and women to reach their potential will serve the dual American ideals of equity *and* excellence in education. The decisive action taken years ago in the wake of Sputnik created a legacy guaranteeing that today's generation would live in a more prosperous and secure society than that of their predecessors. It is our collective responsibility today to do the same, and ensure that future generations reap the benefits of our choice to act. We believe that the recommendations set forth in this report represent one step of many towards continuing this legacy.

INTRODUCTION

In 1957, under the shadow cast by the Soviet Union's successful launch of Sputnik, the United States embarked on a coordinated, decade-long effort to recruit and educate the "best and brightest" who subsequently would form a new generation of leaders and innovators in science and engineering (S&E). This endeavor ushered in a new era of unprecedented scientific and technological advancement in the Nation, leading to the creation of new industries and job opportunities, improvements in national security, and enhancements in our quality of life. At the root of this progress was a nationwide focus on excellence in science, technology, engineering, and mathematics (STEM) education and talent development, along with a substantial investment in research and development (R&D). By the 1970s, however, this national sense of urgency and commitment to excellence in STEM education had lapsed into complacency. In 1983, the landmark report, *A Nation at Risk*, noted that "the ideal of academic excellence as the primary goal of schooling seems to be fading across the board in American education."[3] In 2005, nearly a quarter century after *A Nation at Risk*, the alarm once again was sounded over the looming challenge to U.S. pre-eminence in science and technology (S&T) in the National Academies' seminal report, *Rising Above the Gathering Storm*.[4] This report posited that in the 21st century, educated, talented, motivated people and their ideas are paramount to creating the innovations that will sustain America's prosperity.[5] Finally, in 2009, the Administration's *Strategy for American Innovation* argued for investing in the building blocks of innovation, promoting competitive markets, and catalyzing breakthroughs for our Nation's priorities.

A critical facet of America's historical advantage in S&T innovation has been the ability to attract, develop, and retain talented individuals from abroad. Indeed, over the past few decades, many STEM fields in the United States have become increasingly dependent on foreign-born talent. However, global competition for STEM talent is growing as many countries increase their R&D capacity and improve their own STEM education systems. In light of this, it remains essential that the Nation not only continue to attract STEM talent from abroad, but also renew and redouble its efforts to identify and develop domestic human capital as well.

The Board's 2-year examination of this issue made clear one fundamental reality: ***the U.S. education system too frequently fails to identify and develop our most talented and motivated students who will become the next generation of innovators***. Whether this group of students has access to appropriate resources seems to be an accident of birth—whether they are a part of a supportive and knowledgeable family or are residing in a community that has programs and opportunities available to them. There are students in every demographic and in every school district in the United States with enormous potential to become our future STEM leaders and to define the leading edge of scientific discovery and technological innovation. Some of our Nation's most talented students—perhaps through sheer individual will, good fortune, and circumstance—rise through the educational system and become leading contributors to the scientific workforce. Regrettably, far too many of our most able students are neither discovered nor developed, particularly those who have not had adequate access to educational resources, have not been inspired to pursue STEM, or who have faced numerous other barriers to achievement. The possibility of reaching one's potential should not be met with ambivalence, left to chance, or limited to those with financial means. Rather, the opportunity for excellence is a fundamental American value and should be afforded to all.

The possibility of reaching one's potential should not be met with ambivalence, left to chance, or limited to those with financial means. Rather, the opportunity for excellence is a fundamental American value and should be afforded to all.

Although many past and current educational reforms have focused on the vital goal of raising the general performance of all students, far fewer have focused on raising the ceiling of achievement for our Nation's most talented and motivated students. The Board asserts that educational opportunity is not a zero-sum game: *true equity* means we must address the needs of *all* students. Mutually reinforcing results can be realized when we improve general educational performance as well as identify and stimulate potential leaders in STEM whose creativity and ideas can benefit all. The critical goal of increasing STEM proficiency and general scientific literacy does not compete with, but rather complements, today's renewed clarion call for excellence. The needed focus on excellent STEM instruction that will inspire and excite those who might pursue STEM careers is crucial for *all* learners.

Today, on the 60th anniversary of the National Science Foundation, the United States is confronted with a clear choice between action and complacency. The Board firmly believes that a coherent, proactive, sustained effort to identify and develop our Nation's future STEM innovators will help drive future economic prosperity, improve the quality of life for all, and ensure both equity *and* excellence in education.

STEM "innovators" are defined in this report as those individuals who have developed the expertise to become leading STEM professionals and perhaps the creators of significant breakthroughs or advances in scientific and technological understanding. Historical examples include Edison, Ford, Fleming, Pasteur, Einstein, and Curie. This report alternately refers to the children and young adults who have the most potential to become STEM innovators as "talented and motivated" or "high-ability" or "gifted."[6] Their capabilities often include mathematical and spatial abilities[7] alone or in combination with verbal aptitude, along with other factors such as creativity, leadership, self-motivation, and a diligent work ethic. In an increasingly technological society, innovation is frequently an interdisciplinary endeavor and many traditional non-STEM fields require scientific, spatial, and quantitative talents.

RATIONALE

Two sets of fundamental national values and needs underlie the findings and recommendations proffered in this report. The first set relates to the national need for the entire Country to reap the full rewards of science and technology and their application. America has benefited tremendously over the past 60 years from its investments in developing the world's top scientific enterprise. Increased efforts in a variety of areas, including development of our human capital, will be required to maintain America's international position in S&T as other countries have recognized this accomplishment and seek to emulate it. The second set relates to the American value reflected in providing equal opportunities for all students to reach their full potential and thrive in modern society. Serving the needs of all students, including high-ability students, will help achieve our Country's aspiration for true equality of educational opportunity and will facilitate the development of the innovators of tomorrow who can lead the way forward. Our combined actions today towards meeting these two values and needs will serve as our legacy to the next generation.

Developing Future STEM Innovators: An Economic Imperative

The identification and development of our Nation's human capital are vital to creating new jobs, improving our quality of life, and maintaining our position as a global leader in S&T. In 1945, in the immediate aftermath of World War II, and long before "innovation" became commonplace in our collective vernacular, the necessity of progress in STEM fields was emerging. In *Science–The Endless Frontier*, Vannevar Bush wrote:

> *Our hope is that there will be full employment, and that the production of goods and services will serve to raise our standard of living... Surely we will not get there by standing still... There must be a stream of new scientific knowledge to turn the wheels of private and public enterprise.*[8]

Since then, and with increasing frequency over the past decade, a variety of prominent government and private organizations have warned against "standing still" and forcefully articulated the importance of innovation and talent development to the U.S. economy. Indeed, the innovations that spawn high-technology industries will create new employment opportunities at a rate that exceeds traditional manufacturing industries.[9] A full 65 years since the publication of *Science–The Endless Frontier*, and as the world recovers from a global economic downturn, the unmistakable link between the prescient words of Vannevar Bush in 1945 and those of the Administration in 2010 engenders a renewed poignancy:

> *In our increasingly interconnected and globally competitive world economy, unleashing innovation is an essential component of a comprehensive economic strategy. As global competition erodes the return to traditional practices, the key to developing more jobs and more prosperity will be to create and deploy new products and processes.*[10]

Innovation is the complex process of introducing novel ideas into use or practice in order to develop cutting-edge breakthroughs in emergent fields (e.g., energy sustainability, personalized medicine) as well as novel solutions to age-old problems (e.g., the need for clean and abundant water). Innovation requires highly able, determined, and creative leaders and thinkers. We are now living in what the Council on Competitiveness calls the "conceptual economy," where competitive advantage and value creation rely on "insight, imagination, and ingenuity."[11] So, where will we find our future STEM innovators? Longitudinal data show that intellectually talented individuals who can be identified at an early age (and then supported in their learning) generate a disproportionate number of Fortune 500 patents, peer-reviewed STEM publications, and other creative achievements, and comprise a disproportionate number of tenured academic faculty at top universities.[12] Clearly then, a critical challenge in this "conceptual economy" is to discover and then develop the next generation of innovators who will help create the "products and processes" that will fuel our future economic prosperity.

Is the United States meeting this challenge in a world where international competition is accelerating? Although the United States remains among the leaders in key metrics of innovation and R&D investment, there may be cause for concern. Distressingly, students in other countries are outperforming even our highest-achieving students. In the 2006 Program for International Student Assessment (PISA) test, U.S. 15-year-olds in the 90th percentile (our top students) scored below their peers in 29 countries on mathematics literacy, and below 12 countries on science literacy.[13] Similarly, 6 percent of America's eighth graders reached the advanced benchmark in mathematics on the 2007 Trends in International Mathematics and Science Study (TIMSS).[14] Though this marks an above average score, the performance of U.S. students fell well behind several key competitors. For example, 40 percent of eighth graders from the Republic of Korea and Singapore, and 45 percent of eighth graders from Chinese Taipei (Taiwan) reached the advanced benchmark in mathematics.[15]

Some worrisome indicators are also present beyond K-12 within the higher education system and STEM workforce. There has been an ongoing debate among experts whether there are indeed deficiencies in the U.S. STEM pipeline that inevitably will lead to a future workforce shortage in at least some S&E fields. Critically, even if the overall supply of U.S. citizens entering the STEM pipeline is equal to the demand (or even exceeds demand in some fields), there is evidence that top U.S. students, who have a disproportionate potential to become future innovators, are eschewing careers in S&E. A 2002 analysis showed that between 1992 and 2000, the number of the highest- achieving students intending to enter graduate study in an S&E field declined 8 percent overall, with particularly steep declines in engineering (25 percent) and mathematics (19 percent).[16] Similarly, a more recent report provided evidence that, between the 1 990s and mid-2000s, there was a sharp decline in the number of highest-achieving U.S. high school graduates enrolling in or completing a STEM major in college.[17] While the percentage of top U.S. students entering many S&E fields has declined in recent years, many of these same fields have become increasingly reliant on foreign-born talent.[18] For example:

- Compared to their U.S. counterparts, undergraduate students in foreign countries chose natural science and engineering (NS &E) disciplines as their primary field of study at considerably higher rates. According to the most recent data, 25 percent of undergraduates in the European Union, 47 percent in China, and 38 percent in South Korea chose an NS&E major, compared to only 16 percent of U.S. undergraduates.[19]
- This trend continues further along the STEM pipeline: 33 percent of all U.S. STEM doctoral students in U.S. universities are foreign students on temporary visas, and 57 percent of U.S. postdoctoral fellows in STEM fields hold temporary visas.[20]
- Foreign-born doctoral degree holders constitute an increasing share of the S&E workforce. In 2003, foreign-born doctorate holders represented about half of the workforce in engineering and computer science, and 37 percent and 43 percent of the workers in the physical sciences and mathematics, respectively.[21]

Attracting and retaining foreign-born talent remains an essential pillar of our Nation's STEM enterprise. As global demand for STEM talent surges, we cannot reliably expect that the best and brightest from abroad will remain in the United States and continue to be a sufficient source of talent. It is essential that we develop our own domestic human capital as well. Ideally, foreign talent should augment a robust domestic STEM talent pipeline, not compensate for its deficiencies.

Our Nation's success in developing future STEM innovators rests squarely on the capacity of our education system to identify and nurture ability. This ability can manifest itself in a variety of ways, across many different developmental stages. In the United States, assessments of verbal and mathematical aptitude are well-established and widely used. Yet, a talent highly valuable for developing STEM excellence—spatial ability—is not measured and hence missed. Recent research indicates that current mathematics and verbal talent assessments would miss 70 percent of students scoring in the top 1 percent of spatial ability.[22] Individuals with such talents constitute a lost resource for creating future STEM innovation, since 90 percent of STEM doctorate holders scored in the top quartile of spatial ability during adolescence.[23] As discussed in the next section, another unrealized resource is young men and women from lower-income backgrounds and minorities traditionally underrepresented in STEM. Underrepresented minorities are disproportionately absent from the current STEM workforce but comprise the fastest growing college-aged population in the United States.[24] Similarly, though our Country has made laudable strides in narrowing the gender gap in STEM participation, women are still underrepresented in fields such as engineering, computer science, and the physical sciences.[25]

While the need for future STEM leaders and visionaries is great, our Nation sits atop an untapped talent gold mine. We are faced, therefore, with a clear and profound choice between action and complacency. We believe the choice is as simple as it is vital: Securing our Nation's continued economic prosperity will require the proactive identification and development of talented young men and women from *all* demographics with *all* types of STEM-related abilities who have the potential to become our next generation of STEM innovators.

In light of the economic importance of scientific and technological innovation, increasing global competition, and our dependence on foreign-born talent, we must reawaken a national expectation of excellence. The 2005 Business Roundtable report, *Tapping America's Potential: The Education for Innovation Initiative*, effectively enunciates this point:

> *One of the pillars of American economic prosperity—our scientific and technological superiority—is beginning to atrophy even as other nations are developing their own human capital. If we wait for a dramatic event—a 21st-century version of Sputnik—it will be too late. There may be no attack, no moment of epiphany, no catastrophe that will suddenly demonstrate the threat. Rather, there will be a slow withering, a gradual decline, a widening gap between a complacent America and countries with the drive, commitment and vision to take our place.*[26]

Our continued economic prosperity will depend on a skilled workforce, particularly at the leading edge of science and technology, where a diverse legion of creative, motivated and innovative individuals is essential. Changing course is a long-term proposition. It requires significant foresight and early intervention. Although mastery of a STEM discipline requires over a decade of intensive study after high school, the interest (or disinterest) in STEM germinates early in K-12, maybe even in early childhood. Likewise, no matter how talented the individual, realization of this potential may not occur on its own. Development of our Country's human capital requires the identification and development of all types of STEM-related talents, the encouragement of intellectual ambition, an anticipation of excellence rather than simply competency, and the sustained nurturing of the creative spark. America can create, through its educational system, the next generation of preeminent scientists, engineers, inventors, and entrepreneurs when it focuses its collective will on that critical goal. As a society, we will reap the benefits for decades to come.

Opportunity for Excellence: A Fundamental Value

Equality in educational opportunity means that *all* students fundamentally deserve the chance to succeed in reaching their highest potential. When disparities in academic achievement exist among populations, we have marshaled our collective will in an effort to narrow these gaps. As a result of these efforts, the U.S. education system has experienced some notable improvements. Nonetheless, too many students in America are not achieving even at modest levels, and great disparities continue to exist in the quality of education afforded to students around the Nation. Efforts to raise the educational achievement for all students must not only be continued in earnest, but accelerated. However, to reach *true* equality of opportunity, and to ensure that potential does not go unrealized, we must not overlook the educational needs of our Nation's most talented and motivated students. Too often, U.S. students with tremendous potential to become our future innovators go unrecognized and undeveloped. The dual goals of raising the floor of base-level performance and elevating the ceiling for achievement are not mutually exclusive. The Board believes that both equity *and* excellence are not only possible and mutually reinforcing, but necessary to achieve the American ideal as eloquently articulated 65 years ago in *Science–The Endless Frontier*:

> *We think it is much the best plan, in this constitutional Republic, that opportunity be held out to all kinds and conditions of men [and women] whereby they can better themselves. This is the American way, this is the way the United States has become what it is. We think it is very important that circumstances be such that there be no ceilings, other than ability itself, to intellectual ambition. We think it very important that...if he [and she] has what it takes, the sky is the limit.*[27]

Unfortunately, individuals with a high level of ability and determination frequently lack the opportunities needed to reach their potential. There are examples of successful programs and interventions aimed at advanced learners. Many of today's top scientists, inventors, and entrepreneurs participated in one or more of these programs at some point in their academic development.[28] Indeed, data show that a high density of advanced pre-collegiate learning opportunities among mathematically talented youth has been linked to subsequent accomplishments in STEM.[29] Yet the scale of these programs is often small, and access to these programs is frequently limited. More often than not, across the education ecosystem, we see a patchwork of individual, often *ad hoc* provisions implemented and funded at the local level; these approaches have been instrumental for many of today's STEM innovators and should continue. In addition, a coherent, long-term, state- or Nation-wide plan to develop the next generation of leaders in STEM is also needed. Our Nation has too often left to chance the fate of those with exceptional talent rather than ensuring widespread, systematic, and appropriate opportunities to flourish.

Our Nation has too often left to chance the fate of those with exceptional talent rather than ensuring widespread, systematic, and appropriate opportunities to flourish.

Historically and by law, states and local education agencies (LEAs) are the primary source of support for talented learners and often represent the only source of support. Not surprisingly then, the funding and education policies among the states and even districts within the same state vary considerably. The National Association for Gifted Children's (NAGC) *State of the States in Gifted Education* report describes the situation for gifted and talented education at the state and local levels.[30] In 2008-2009, out of the 45 states that fully responded to the NAGC survey:

- 32 states required school districts to provide some services for gifted and talented students. Of these 32 states, only 6 reported fully funding these services.
- 12 states provided no funds to support gifted education.
- Among the 14 states that reported both funding levels and numbers of identified gifted students, the yearly allocation per child ranged from less than 2 dollars to approximately 760 dollars.
- 11 states required districts to accept gifted identification decisions from other districts in the same state.
- Most high-ability children were placed in the general classroom where the majority of teachers have little or no specialized training in working with gifted children.
- Only five states required *all* teachers to have pre-service training in gifted and talented education. Only five states required annual professional development for teachers in specialized gifted and talented programs.
- 21 states reported that they neither monitor nor audit local programs for talented students.

Meanwhile, support from the Federal Government at the elementary, middle, and high school levels for our high-ability youth is minimal. A single program at the Department of Education, the *Jacob K. Javits Gifted and Talented Students Education Program*, is specifically dedicated to supporting talented students. Even this program is routinely targeted for elimination due to Federal budgetary considerations.[31] The National Science Foundation (NSF) has a few general STEM education programs that could potentially support research in this area, such as *Discovery Research K-12* and *Research and Evaluation on Education in Science and Engineering*. However, NSF currently does not have any programs or initiatives specifically dedicated to the direct support of, or research into, our Nation's future innovators at the K- 12 level.[32] In the absence of a coordinated plan and consistent opportunities for young men and women across the entire Country, talented students may slip through the cracks or face bureaucratic, institutional, societal, and/or other hurdles that stymie their progress and suppress intellectual ambition.[33]

Findings from the 2007 *Achievement Trap* report suggest that educators, policy-makers and even parents erroneously assume that high-achieving young men and women will continue to achieve at high levels on their own and do not need additional support.[34] However, an analysis of the performance of high-achieving students on the National Assessment of Educational Progress (NAEP) paints a different picture. Whereas the scores of students in the tenth percentile (low achievers) have seen significant improvements over the past decade, test scores for students in the ninetieth percentile (high achievers) have, at best, experienced modest gains and, at worst, stagnated.[35] The situation for highly talented and motivated students from lower socioeconomic backgrounds and underrepresented minorities is especially alarming. Although high achievers can be found in every geographic area and every race/ethnicity,[36] data from *Achievement Trap* show that talented students from lower-income levels are underrepresented and lose ground at virtually every stage along the educational continuum.[37] For example:

- Approximately 3.4 million children rank in the top quartile academically and come from households with incomes below the national median. At every educational level they fare worse in terms of academic outcomes compared to their higher-income counterparts. Thus, these students represent a potentially significant source of underdeveloped talent.
- Disparities begin early: In first grade, 72 percent of students in the top quarter of their class come from higher-income families, compared to only 28 percent of lower-income students. According to the authors of *Achievement Trap*, this disparity means, "200,000 or more children from lower-income backgrounds appear to be lost each year from the ranks of high achievers before their formal education ever begins."[38]
- Lower-income students *fall out of* the top quartile during elementary and high school at significantly higher rates than their higher-income peers.
- Lower-income students are considerably less likely than higher-income students to *rise into* the top quartile during elementary and high school, attend the most selective colleges, finish a baccalaureate degree, or go on to complete a graduate degree.

Likewise, achievement gaps between white or Asian/Pacific Islander students and minorities traditionally underrepresented in STEM exist at *all* levels, including significant gaps among the highest-performing students. For example, a recent analysis of both NAEP and state assessment data shows that large achievement gaps in mathematics performance continue to persist between white and underrepresented minority high achievers.[39] Moreover, extremely low percentages of minority students reach the advanced level on NAEP. In 2007, only 0.8 percent of African American students and 1.5 percent of Hispanic students reached the advanced level on the fourth grade NAEP mathematics exam. Similarly, only 0.9 percent of African American students and 1.8 percent of Hispanic students reached the advanced level on the eighth grade NAEP mathematics exam. In comparison, 7.6 percent and 9.4 percent of white students reached the advanced level in mathematics in fourth and eighth grade, respectively.[40] These and other data underscore a systematic lack of opportunities and support for underrepresented minority students, inadequate teaching, and an absence of both real-life, hands-on experiences with STEM materials and positive role models of STEM professionals.[41]

Talented underrepresented minorities also face significant inequities that contribute to the achievement gap at the high end of academic performance. For instance, African American, Hispanic and American Indian/Native Alaskan students are underrepresented in gifted and talented programs in K-12, attain lower SAT scores, are less likely to take advanced mathematics courses or Advanced Placement (AP) exams, attend less prestigious higher education institutions, and are less likely to graduate with a degree compared to whites and Asians/Pacific Islanders.[42] Consequently, many of our most talented and determined lower-income students and minority students traditionally underrepresented in STEM are never identified or given an equal opportunity to realize their enormous potential, and, therefore, constitute a considerable source of untapped talent.

The attitudes of educators, policy-makers, parents, students' peers, and even students themselves toward excellence can act as facilitators of, or form considerable barriers to, academic achievement. Our society almost universally applauds certain areas of talent—sports in particular, and to a lesser extent music and the arts. In contrast, intellectual talent often generates attitudes ranging from ambivalence to outright hostility.[44] A 2008 survey revealed that 86 percent of teachers said that to attain the American ideals of justice and equality, it was important to focus equally on all students, regardless of their backgrounds or achievement levels. Nevertheless, only 23 percent of teachers indicated that academically advanced students were a priority at their school. Similarly, 73 percent of teachers surveyed agreed that "too often, the brightest students are bored and under-challenged in school" and are not given sufficient opportunities to thrive.[45] Moreover, the vast majority of general education teachers receive little or no training on pedagogical best practices for talented learners.[46] Consequently, most teachers make only minor and irregular modifications to the regular classroom curriculum to serve the academic needs of these students.[47] With the proper attention from teachers and administrators, these students could access a component of the learning support infrastructure vital to achievement.

> *"If you win the NCAA championship, you come to the White House. Well, if you're a young person and you've produced the best experiment or design, the best hardware or software, you ought to be recognized for that achievement, too. Scientists and engineers ought to stand side by side with athletes and entertainers as role models."*
>
> Barack Obama,
> U.S. President
> November 23, 2009

An unsupportive or negative learning ecosystem can undermine self-efficacy—that is, beliefs about one's capabilities to learn or perform at designated levels. Women, for example, are underrepresented in the engineering profession, and female engineering undergraduates experience high attrition rates.[48] Low self-efficacy of aspiring female engineers correlates with a perceived lack of inclusion in engineering classrooms, possibly due to negative attitudes of peers and faculty, and could be partly to blame for this phenomenon.[49] Some evidence exists that the low participation of underrepresented minorities in gifted education programs is caused in part by the diminished expectations of educators, due to negative and stereotypic views regarding the academic ability of these students.[50] The resulting lack of diversity then may lead underrepresented minority students who *do* participate in gifted programs to feel isolated and not part of the classroom experience. As a result, they may withdraw from classroom activities and hide their abilities from teachers and peers.[51]

Similarly, misconceptions regarding intelligence may form additional barriers to achievement. Some research indicates that, to a certain extent, ability and intelligence are malleable—that is, rather than being a fixed trait, some abilities potentially can be developed over time with hard work and the proper support. Nevertheless, many students and teachers believe that intelligence is a fixed trait, and this belief can hinder a student's motivation and ability to learn and improve. This mindset can have particularly pernicious effects on the learning and achievement of women and underrepresented minorities.[52] Critically, the belief in fixed intelligence can create a fear of failure within a student, yet innovation requires risk taking and outside-of-the-box thinking. Even studies of geniuses and landmark creative accomplishments demonstrate that talent alone is insufficient, and sustained, diligent training and practice are indispensable.[53]

Further compounding the issue, talented students from many demographics may face a social stigma that too often accompanies academic success. In particular, African American and Hispanic students in public schools sometimes experience anti-intellectual social pressure from peers, which negatively impacts performance.[54] High-ability Hispanic students also may face linguistic barriers that hinder academic achievement, and tend to have lower academic aspirations compared to higher- income, white students even when they possess the requisite ability and training.[55] Because women and underrepresented minorities are disproportionately absent from many STEM disciplines, and Hispanics in particular are the largest growing college-aged population,[56] identifying and supporting these students are vital to both the concept of equity and ensuring a robust, diverse workforce for the future. Creating a society and culture whose institutions, especially schools, value and reward academic excellence represents a national responsibility.

RECOMMENDATIONS

The innovation continuum, from identification and development of talented and creative individuals through the education system, to a STEM career, and then to major scientific breakthroughs or to the creation of a novel product, is both vast and complex. Even with unlimited time and resources, the Board would be hard-pressed to address every facet of the innovation process. Therefore, we have chosen to focus on the human capital component, especially early in the education system, where we feel much of our domestic talent goes unrecognized and undeveloped. Thus, the following recommendations are not exhaustive and not intended to be the final word on the subject.

Though we are focused on identifying and developing future innovators in STEM, several of our recommendations could benefit *all* students.[57] This is by design. Similarly, many findings and techniques found effective for the general population of students may prove useful for high-ability students. The Board recognizes that excellence is an objective that *all* U.S. students should endeavor to attain. Other recommendations we propose clearly reflect our finding that talented students have some learning needs that are distinct from those of the general population. **Ultimately, our hope is that the needs of our future STEM innovators increasingly will become part of the national education discourse among the**

public and policy-makers alike. We encourage others to join our call for a renewed commitment to both equity *and* excellence for all students, so that potential is never squandered, intellectual ambition need not hide, and creative ability blossoms.

In this section, we outline three broad ***keystone recommendations*** based on the findings from the expert panel discussion hosted by the Board in August 2009 (see Appendix III) and the Board's 2-year study of this issue. Contained within each keystone recommendation are multiple specific ***policy actions*** for NSF, the Federal Government, and/or the Nation. Following the policy actions, we propose a ***research agenda*** for each keystone recommendation. Though a substantial body of research and considerable discussion with experts informs our policy recommendations, much remains unknown. NSF, through its broad investments in STEM education, the learning sciences, workforce development, and STEM research, is well positioned to facilitate both a nuanced examination of human capital development and a high-level survey of the entire innovation ecosystem. These research findings will inform policy-making in critical areas, such as identifying future innovators and improving teaching effectiveness, to maximize long-term returns on our investment.

I. Provide Opportunities for Excellence

Improve the access to and availability of effective K-12 formal and informal education programs and interventions to meet the needs of future STEM innovators.

Findings

Inconsistent efforts and resources have been expended to support our Nation's most talented and determined students. As a result, their educational needs often go unmet. Experience shows that without a widespread, equitable, and coherent support system, a full realization of potential is unlikely. Policy-makers have made notable efforts over the past decade to implement standards and accountability in the U.S. education system (e.g., *No Child Left Behind Act of 2001* (NCLB)). Unfortunately, these standards have not mandated increases in—or even the measurement of—advanced levels of educational performance. Regardless of the rigor of standards, schools have become focused on getting children across the basic proficiency threshold.[58] Currently, there are no "standards of excellence" to which schools are held.

In America, it should be possible, even essential, to elevate the achievement of low-performing at- risk groups while simultaneously lifting the ceiling of achievement for our future innovators.

There are many individual successful programs available for talented and motivated students interested in STEM.[59] However, many of the existing opportunities are limited in scope and access, or suffer from a lack of resources. In America, it should be possible, even essential, to elevate the achievement of low-performing at-risk groups while simultaneously lifting the ceiling of achievement for our future innovators. Consistent, coherent, and coordinated opportunities that challenge and inspire high- potential students both in and out of the classroom are needed. The Board believes the following findings inform and support the policy actions and research agenda proposed below.

- Talented, motivated students tend to master curriculum content at a rapid rate and often have mastered 40 to 50 percent of grade-level content before entering each grade.[60] This hunger for new information and further learning can quickly atrophy into boredom if not satisfied.[61] Increased classroom "time on task" is an idea that is gaining popularity among policy-makers, but time on task is squandered if it is spent on a subject that a student has already mastered. Therefore, these students require classroom content and pacing suitable to their individual learning styles, interests, and abilities.[62]
- Research shows that curriculum acceleration, or accelerative learning, is one of the most effective interventions for high-ability individuals.[63] Acceleration is an intervention that moves students through a standard school curriculum faster, or at younger ages, than typical without necessarily requiring the development of specialized curricula. The level and pace of a curriculum is synchronized to the intellectual readiness, emotional maturity, and motivation of the student. Research shows that, by-and-large, those students permitted to accelerate not only achieve more but also are happier than those who are not.[64]
- Accelerative learning generally costs very little, but requires school administrative flexibility, particularly for younger students who are more likely to be denied this opportunity by states and LEAs.[65] Similarly, due to bureaucratic hurdles and/or state and local policies, the ability and prior achievement level of students moving or transitioning (e.g., moving from elementary to middle school) to different schools are sometimes ignored, forcing these students to retake coursework they already have mastered.
- In the STEM areas, all students, including the most talented, should have the opportunity to experience inquiry-based learning, peer collaboration, open-ended, real-world problem solving, hands-on training, and interactions with practicing scientists, engineers and other experts.[66] Currently, many of the opportunities for these activities materialize in the form of informal, out-of-school enrichment activities (e.g., summer camps, competitions, science museum visits, Math Circles), rather than as an integrated ingredient of a STEM curriculum. Out-of-school enrichment is extremely valuable, particularly to inspire interest in STEM, but insufficient by itself. Students spend the vast majority of their time in the regular, formal classroom.[67] Formal and informal education are mutually reinforcing and are most effective when synchronized.[68]
- Formal and informal enrichment programs are limited in their prevalence and persistence, particularly for students in poorly funded locales.[69] When combined with acceleration, enrichment is especially powerful and should be included.[70] Emerging technologies can be instrumental in providing schools access to meaningfully enriching STEM resources. Through the Internet students can connect to formal and informal learning opportunities and STEM experts, gain interactive access to world-renowned museum collections and a vast array of digital STEM content, and participate in virtual laboratories.

- Early exposure to STEM is particularly important, since interest in STEM often begins to blossom in elementary school, and early exposure to science can strongly influence future career plans.[71]
- Engineering is a field critical to innovation, and exposure to engineering activities (e.g., robotics and invention competitions) can spark further interest in STEM. However, exposure to engineering at the pre-collegiate level is exceedingly rare.
- Some students who show potential for high achievement are not prepared for advanced content because they have not had access to appropriate resources or have not been challenged by their learning environment. One way to address this issue is through special "bridge" programming. Bridge programs can help elevate student achievement to a level commensurate with individual potential, improve confidence, and enable students to engage in classroom activities at the level of their high-achieving peers so they can fully benefit from the experience.[72]

Policy Actions

A. Encourage states and/or local education agencies to adopt consistent and appropriate policies on differentiated instruction, curriculum acceleration, and enrichment, and to recognize the achievement levels of students moving or transitioning to different schools. States and LEAs should examine ways to remove bureaucratic or policy-related barriers and increase administrative flexibility so as to allow students—beginning in early grades—to proceed at a pace that matches their individual ability and interest.[73]
B. Increase access to and quality of college-level, dual enrollment, and other accelerated coursework (e.g., Advanced Placement and International Baccalaureate classes), as well as high-quality enrichment programs. Particular attention should be given to increasing the participation and success of underrepresented and low-income groups in these classes.
C. Support rigorous, research-based STEM preparation for teachers, particularly general education teachers, who have the most contact with potential STEM innovators at young ages. Attention should be given to training teachers in the most effective methods of teaching STEM content, including hands-on and unstructured problem solving and inquiry-based learning.
D. Provide Federal support to formal and informal programs that have a proven record of accomplishment in stimulating potential STEM innovators. These should include formal education programs that use innovative teaching methods or employ inquiry-based learning, and informal programs, such as robotics and invention competitions, Math Circles, hands-on "lab days", mentoring opportunities, and science fairs. Attention also should be given to programs that satisfy one or more of the following:
   - Provide "bridge" experiences for students with high potential who have not had consistent past opportunities for achievement;
   - Have proven effective in promoting diversity, or reducing the achievement gaps at the "high end" of academic performance based on race/ethnicity, gender, and/or income level;

- Have demonstrated success in lowering the attrition rate at the high school to higher education transition point.

E. Leverage NSF's *Broader Impacts Criterion* to encourage large-scale, sustained partnerships among higher education institutions, museums, industry, content developers and providers, research laboratories and centers, and elementary, middle, and high schools to deploy the Nation's science assets in ways that engage tomorrow's STEM innovators. Particular attention should be given to mentoring opportunities for students in K-12 and partnerships that engage students and teachers in K-12 in entrepreneurial, innovative environments.

F. Create NSF programs that offer portable, merit-based scholarships for talented middle and high school students to participate in challenging enrichment activities, such as summer programs, Math Circles, hands-on research experiences, and competitions. The scholarship criteria should emphasize identifying students who possess high potential but who have not had consistent prior opportunities to demonstrate academic excellence.

G. Increase the technological capabilities and network infrastructure in rural and low-income areas, and expand cyber-learning opportunities. Some examples of these opportunities include access to digital resources, remote connections with STEM experts, the creation of online learning communities, and virtual laboratories.

H. Create a national database of formal and informal education opportunities for highly talented students, and publicize and promote such opportunities nationally to parents, education professionals, and content and resource providers.

Research Agenda

Rigorous evaluation data regarding existing educational services and their outcomes are frequently lacking. Therefore, a key component of a research agenda must be a candid analysis of which educational and enrichment interventions work (and how well, for whom, and under what conditions) and which do not, in the short-run and long-term. In a climate where education resources are scarce, it is essential to provide policy-makers with empirical evaluation data to aid their funding decisions. Moreover, evaluation data are equally important for educators and parents who bear the primary responsibility for ensuring that talented children and young adults are given worthwhile opportunities to cultivate their abilities. Some outcome evaluations are available, but few are effective in providing the education community with the generalizable knowledge needed to build better interventions. Although programmatic knowledge—that is, specific information applicable *only* to a particular intervention—is an important component of evaluation, there is a need for generalizable knowledge, which can be used across programs and perhaps even across disciplines. We recommend the following four priorities for the research agenda:

1. Examine NSF's current *Broader Impacts Criterion* relative to STEM education for highly talented and motivated K-12 students. Serving highly talented and motivated individuals in K-12 and beyond should be a means for satisfying this criterion. Higher education institutions are well suited for this role and should be encouraged to do so.

2. Provide support for independent external evaluations on the short- and long-term outcomes of Federal, state, and local programs focused on high-ability individuals or groups. Evaluations should be designed such that data generated are generalizable to a broad array of programs, thus increasing the knowledge base of best practices. Emphasis should be given to studying the impact of these programs on the three criteria listed under "Policy Action D" above.
3. Investigate the scalability and replicability of successful programs and best practices.
4. Support research on designing novel, innovative methods for developing talents. In addition, researchers should explore effective means for implementing these techniques in education schools, and teacher preparation and professional development programs.

II. Cast a Wide Net

Improve the identification of potential STEM innovators—especially among underrepresented populations—by augmenting teacher training and using talent assessments that 1) span the entire K-12 continuum, 2) reflect the multiple domains of ability (e.g., verbal, mathematical, spatial), 3) have sufficient range at the top to distinguish high from extremely high ability, and 4) are appropriately matched to the background, education history, and prior achievement history of the individual.

Findings

The abilities of large numbers of potential future STEM innovators currently go unrecognized and are underdeveloped. Though cognitive ability is only one of many attributes of a future innovator, it is important. Identifying this ability as early as possible is critical for developing an appropriate educational intervention. Abilities may develop at different rates for different individuals, so educators must diligently seek out potential throughout the entire educational continuum. Identification of the most talented students, whether their talents are verbal, mathematical, or spatial, is improved by the use of above-level tests—that is, tests designed for older students—as part of the suite of identification activities. When age-appropriate tests are used, both high- and extremely high-ability students are not distinguishable in the test results. Above-level testing provides vital information that allows for a better tailoring of educational experiences. Likewise, educators require the training and experience to recognize talented students and facilitate their entry into the appropriate programs or interventions.

Research shows that high ability is present across all demographics. Yet, underrepresented minorities and students from low-income backgrounds are disproportionately absent from gifted classrooms and drop out of the "high achieving" category during elementary and high school at alarming rates.[74] Increased global competition for talent and shifting racial/ethnic demographics in the United States underscore the importance of casting a wider net to capture all forms of STEM-related talents, in all of the places it can be found. The policy actions and research agenda below are supported by the following findings:

- Talent is not a binary phenomenon (i.e., "you have it or you don't"). Research shows that ability is dynamic, can be developed over time with proper training, and the developmental process can occur at different paces for different people.[75] Assessments must be given early and often throughout K- 12 rather than a single test administered at a single or just a few developmental stages.
- Identification and development go hand-in-hand. To properly identify and assess students with high potential, interest and talent in STEM should be developed at an early age. Opportunities for educational development in STEM can unmask or improve general or domain-specific cognitive abilities and critical non-cognitive abilities, such as persistence, creativity, and leadership. As these abilities are developed, identification mechanisms improve.
- Verbal and quantitative/mathematical skills are two of the most commonly understood and assessed abilities. Numerous tests for these skills are deployed to identify children whose performance is well beyond that of the typical child. Talent searches are widespread in seventh and eighth grade but less well developed at younger ages.
- Future achievement in STEM is linked to the pattern of abilities present in an individual. For example, mathematical and spatial ability alone or in combination have been associated with STEM expertise and are predictive of a future career in S &E.[76]
- Spatially talented students may not fit the classic model of what parents, the public, and even educators think of as "gifted." Rather than excelling in a typical classroom, these individuals might actively engage in vocational or career training classes or in projects outside of school where they can perform hands-on activities in three dimensions. These students may gravitate to engineering classes if offered early in the curriculum. Individuals with spatial abilities are routinely overlooked because these abilities are rarely measured and, if they are, the results often are not given the proper attention. This is an untapped pool of talent critical for our highly technological society.[77]
- Opportunities for achievement and success have not been afforded equally to all talented individuals or groups. Results from any testing, whether it is on-level or above-level, need to be considered in light of the backgrounds of the students taking the test and their prior opportunities to learn and achieve.[78] A student may not appear exceptional if his or her performance is compared to national norms. However, if individual context, such as being an English language learner, being the first in his or her family to graduate from college or even high school, coming from a low-income background and/or an environment lacking intellectual stimulation is considered, his or her performance may stand out and be indicative of high potential.
- Teachers often act as "gatekeepers" to gifted classrooms and resources.[79] However, they frequently receive inadequate or no training on how to identify and develop students with high potential.[80] The most talented students or students with the highest potential may not always be the "best students" with the highest grades, or the most well behaved students.

Policy Actions

A. Improve pervasiveness and vertical coherence of existing talent assessment systems.
  - Rather than administer a single assessment at a single developmental stage or grade level, provide multiple above-level tests throughout the K-12 continuum.
  - Encourage schools to improve vertical coherence by tracking the progress of students identified as having high ability beginning in kindergarten all the way through to completion of high school and beyond. It should be a category for which schools are monitored for making progress or *adequate yearly progress* if this concept is continued in Federal laws.

B. Expand existing talent assessment tests and identification strategies to the three primary abilities (quantitative/mathematical, verbal, and spatial) so that spatial talent is not neglected. Talent searches should routinely measure spatial ability.

C. Increase access to appropriate above-level tests and student identification mechanisms, especially in economically disadvantaged urban and rural areas. These tests should use standards that are representative of the local norms and account for the prior learning opportunities of the students assessed.

D. Encourage pre-service education and professional development for education professionals (including teachers, principals, and counselors) in the area of STEM talent identification and development. Education schools and other teacher preparation programs should emphasize teacher preparation in all areas of identification, including spatial ability recognition and the identification of talented underrepresented minorities. Teacher training and professional development must rely on the best available research in these areas and should be aligned with evidence of improvements in student identification and outcomes.

E. Encourage pediatricians and early childhood educators, especially *Head Start* teachers, to become knowledgeable about early signs of talent and the need for its nurturance.

Research Agenda

Much is still unknown about the various forms of ability and their relationship to future innovation. Put simply: How do we best identify individuals who have the potential for future creative and innovative accomplishments in the STEM fields? Clearly, cognitive ability matters, as do non-cognitive factors such as motivation, hard work, and the learning ecosystem (discussed in Recommendation III). However, if the ultimate goal is subsequent sustained innovation, researchers must investigate whether there are other crucial characteristics that are currently overlooked. A properly formed research program will answer this question and elucidate the characteristics that define a future innovator. From this research, development of identification paradigms may be possible, encompassing a mosaic of cognitive and non-cognitive attributes that can help facilitate the discovery of more potential future innovators at an early age. Similarly, this research could shed new light on the role of spatial ability in STEM innovation. Finally, research should emphasize understanding how individual context (e.g., race/ethnicity, gender, socioeconomic status, age, locale) can alter the effectiveness of an identification strategy. We recommend the following three priorities for the research agenda:

1. Support research into identifying the cognitive and non-cognitive characteristics of future STEM innovators. Essential research questions include: What are the relative contributions of cognitive factors, such as domain-specific abilities (e.g., quantitative/mathematical, verbal, spatial), and non-cognitive attributes (e.g., motivation, leadership, resilience, creativity)? Is there a pattern of attributes unique to future STEM innovators, and how can schools teach creativity and innovative thinking?
2. Examine means for developing proper wide-scale assessment systems of all forms of abilities, particularly spatial ability and other overlooked talents (i.e., develop a research-based set of talent identification "best practices").
3. Investigate the optimal strategies to identify underrepresented individuals or groups that have the potential to become future STEM innovators. Particular attention should be given to examining the obstacles to identifying individuals with high potential and developing methods for overcoming them.

III. Foster a Supportive Ecosystem

Enhance the learning infrastructure support system for students by improving educator preparation and encouraging a culture that values academic excellence and innovation in families, local communities, schools, and the Nation.

Findings

Most learning occurs in a social context or ecosystem. This learning ecosystem includes teachers, principals and school administrators, guidance counselors, families, peers, neighborhoods, and a variety of other persons or factors that can assist or thwart academic development. The general attitude of these individuals and groups towards academic excellence can decidedly influence the learning ecosystem. Portions of our society often regard early intellectual achievement with ambivalence, and in some cases, outright hostility. This was not always the case. Equity in education is an empty concept without excellence. To ensure that our Nation continues to thrive in an increasingly competitive global economy, we must renew our efforts to create learning environments that nurture and celebrate intellectual achievement. The following findings support the policy actions and research agenda proposed below:

- Intellectually talented children and young adults can readily detect ambivalence, low expectations, or other negative attitudes within their learning ecosystem. Worse yet, sometimes these students face outright hostility. This often results in adverse consequences, such as poor self-efficacy, loss of motivation, and intellectual regression.[81]
- We should honor all of the gifts of our students, including academic talent, artistic abilities, inventiveness, and athletic accomplishments. In light of our Country's needs at both the national and regional/local levels, encouraging the pursuit of STEM careers by our talented students is particularly essential.

- Teachers are one of the most critical elements in the learning ecosystem. They must be well prepared and enthusiastic—characteristics that are necessary for the education of all students, not just the most talented. However, to teach potential STEM innovators, effective teachers must possess both exceptional subject content mastery and special pedagogical preparation for working with such students. Currently, most teachers receive very little preparation for working with or identifying talented students.[82] Research shows that teachers who are provided with this experience display a more positive attitude towards working with these students, are better skilled at identifying talent, and are more effective educators than those who do not receive such training.[83]

To ensure that our Nation continues to thrive in an increasingly competitive global economy, we must renew our efforts to create a learning environment that nurtures and celebrates intellectual achievement.

- Lack of administrative support, administrative or bureaucratic hurdles, and the absence of a positive school culture can discourage intellectual achievement, and in some cases, lead to students demonstrating regressive, oppositional behavior towards formal education.[84] Low expectations for some students, a lack of school leadership and teacher understanding of student potential and talent, and other negative attitudes and assumptions adversely affect the availability of programs and services for students advanced in the STEM areas. These factors also generate a lack of coherence and vertical alignment in the programming and interventions that *do* exist.
- Parents/guardians have the primary opportunity to instill in their children an expectation of excellence and a strong work ethic. Aversion or fear of certain subjects, such as mathematics (e.g., "math phobia"), is readily passed from teachers to students.[85] It is also likely that these anxieties are passed from parents/guardians to children. Parents/guardians can support future STEM innovators if they display a positive attitude towards learning and discovery to their children at the earliest ages. For instance, it should be just as unacceptable to be poor at math as it would be to be poor at reading.
- Motivation, achievement, and creativity are influenced by peer interactions. Connections with motivated, like-minded, and highly able peers can help foster a positive learning ecosystem and can be highly affirming.[86] Absence of this connection or peer hostility can quickly stifle early intellectual ambition. The Internet enables students to connect to both peers and learning opportunities unbounded by geography.
- Talent development serves both national and regional/local interests. However, resources to support this endeavor are predominantly derived from state and local agencies and possibly other funding sources, such as nonprofit entities.

Policy Actions

A. Create a national campaign aimed at increasing the appreciation of academic excellence and transforming stereotypes towards potential STEM innovators. The campaign should focus on individuals and groups involved in generating a positive learning support ecosystem, including educators, scientists and other STEM professionals, policy-makers, and the media. The media is critical to developing a culture that values STEM excellence. Learning opportunities should be available for parents/guardians about the importance of developing their children's abilities at the earliest ages, supporting their children's academic achievement and creative endeavors, and fostering a family culture that expects excellence. Prior STEM communication efforts, such as the National Academy of Engineering's 2008 report, *Changing the Conversation: Messages for Improving Public Understanding of Engineering*, could provide a useful blueprint for this campaign.[87]

B. Encourage the creation of positive school environments that foster excellence by providing professional development opportunities for teachers, principals, counselors, and other key school staff.
   - For teachers, provide professional development in STEM instructional practices shown to improve achievement, creativity, and motivation among talented students.
   - For principals and other administrators, provide professional development opportunities aimed at strengthening the leadership skills necessary to cultivate a supportive learning ecosystem for both teachers and *all* students.
   - For counselors and other key school staff, provide professional development aimed at understanding the educational needs of talented students from diverse backgrounds and with diverse interests.

   Attention should be given to professional development aimed at transforming negative attitudes and mindsets of educators and students regarding abilities and intelligence, and reversing underachievement in students with high potential.

C. Support the expansion of computing and communications infrastructure in elementary, middle, and high schools to foster peer-to-peer connections and collaborations, and direct connections with the scientific research community.

D. Hold schools, and perhaps districts and states, accountable for the performance of the very top students at each grade. Progress should be monitored for the top 10 percent and top 1 percent of students in each school using assessments that can adequately measure their performance (i.e., assessments must have an appropriately "high ceiling" to measure the full range of performance).[88] Schools and districts that demonstrate progress (e.g., increased student achievement, closing of achievement gaps at the "high end") should be rewarded. Conversely, sanctions should apply if these students are not making progress consistent with their talents and potential, just as it applies for other subgroups of students. We measure what is valued and their performance should be valued as well.

E. Have NSF, in partnership with the Institute of Education Sciences, hold a high-level conference to bring together researchers in the learning sciences, other scientists, education school administrators, current teachers and principals, and teacher professional associations to discuss teacher preparation and pedagogical best practices aimed at fostering innovative thinking in children and in young adults.

Research Agenda

Much work remains to understand fully the role of the learning support ecosystem and its relationship to future innovation. Individual ability and pattern of ability are clearly important as we describe above,[89] yet the development of an innovator does not take place in a vacuum. Instead, innovation occurs within a social context. A supportive learning environment can certainly enhance academic achievement, but more research is required to understand the characteristics of an effective ecosystem that can produce future leaders in STEM. Ability is present across all demographics, and educational opportunities and social context are likely contributors to achievement differences at the high end. Therefore, it is also vital to analyze specific contextual group indicators, such as cultural identity, gender, and socioeconomic status, that may have a disproportionate impact on underrepresented populations in STEM. We recommend the following three priorities for the research agenda:

1. Support research focused on identifying the characteristics of a learning ecosystem that facilitates near-term academic achievement and long-term production of innovation. Cross-cultural studies might be especially useful.
2. Investigate the individual contributions of, and the interplay between, the cognitive and non-cognitive attributes of an individual, and the learning ecosystem, in leading to future development of STEM innovators.
3. Study the impacts of self-perception, cultural identity, and external pressures (e.g., perceptions, stereotypes) on learning and future innovative achievement in STEM. Examine methods to overcome obstacles associated with these factors.

CONCLUSION

In November 1944, as World War II drew to a close, President Franklin Delano Roosevelt wrote in a letter to Vannevar Bush:

> *New frontiers of the mind are before us, and if they are pioneered with the same vision, boldness, and drive with which we have waged this war we can create a fuller and more fruitful employment and a fuller and more fruitful life.*[90]

Today, in the midst of another war and economic struggle, the National Science Board firmly believes that to secure our Nation's long-term prosperity we must identify and develop the talented young men and women who will become the next generation of STEM innovators. This endeavor begins with educational opportunities: the opportunity to achieve to the best of one's ability, the opportunity to think creatively, and the opportunity for the engagement and excitement that STEM provides. The rewards for our collective commitment will be numerous. The Board cannot improve upon the eloquent words of Vannevar Bush in his response to President Roosevelt:

> *The pioneer spirit is still vigorous within this Nation. Science offers a largely unexplored hinterland for the pioneer who has the tools for his task. The rewards of such exploration both for the Nation and the individual are great. Scientific progress is one essential key to our*

security as a nation, to our better health, to more jobs, to a higher standard of living, and to our cultural progress.[91]

The Board firmly believes that the recommendations set forth in this report will help to ensure a legacy of continued prosperity for the United States, and engender a renewed sense of excellence in our education system, benefiting generations to come.

APPENDIX I

NSB-08-82
August 13, 2008
Revised February 4, 2010

CHARGE
COMMITTEE ON EDUCATION AND HUMAN RESOURCES EXPERT PANEL DISCUSSION ON PREPARING THE NEXT GENERATION OF STEM INNOVATORS[92]

Purpose

The National Science Board (Board) Committee on Education and Human Resources (CEH) is charged to undertake a study to fulfill the goal articulated in the Board's *National Science Board National Action Plan for Addressing the Critical Needs of the U.S. Science, Technology, Engineering and Mathematics* (STEM) *Education System* (*NSB-07-114*) to enhance "the Nation's ability to produce a numerate and scientifically and technologically literate society and to increase and improve the STEM education workforce." In approving its STEM Action Plan, the Board recognized that "Strategies for producing the next generation of innovators are not explicitly addressed in this action plan and will require subsequent study."

An *ad hoc* Task Group of CEH will lead the study whose purpose will be to identify strategies for increasing the number of STEM innovators in the next generation, and to develop recommendations for how the National Science Foundation, and possibly other Federal entities, might engage in fostering the development of the next generation of STEM innovators and in conducting rigorous research to better understand this process. As part of its effort, the Board will sponsor a two-day expert panel discussion on this topic and produce a white paper from this expert group with recommendations for consideration by the Board.

Statutory Basis

NATIONAL SCIENCE BOARD (42 U.S.C. Section 1863) SEC. 4(j) (2) The Board shall render to the President and to the Congress reports on specific, individual policy matters related to science and engineering and education in science and engineering, as the Board, the President, or the Congress determines the need for such reports.

Link to National or NSF Policy Objective

The Nation needs both financial resources and STEM talent to drive our highly technological and knowledge-based economy. The Board has argued in a number of its recent policy reports that the United States is too dependent on importing STEM talent from other countries, rather than nurturing a sufficient pool of this talent through our own educational system.[93] Other organizations and entities have also addressed issues related to STEM innovators, including the recent report of the National Mathematics Advisory Panel.[94] The President's American Competitiveness Initiative, the America COMPETES Act legislation, and the National Academies report, *Rising Above the Gathering Storm*, all recognize the importance of STEM talent to our economy. It would be appropriate for NSF, with a mission that encompasses both the development of STEM excellence (e.g., the NSF Graduate Fellowships) and equity (e.g., the Math and Science Partnerships program) to take a lead toward enabling our Nation to make headway on the dual objectives of global economic competitiveness and educational equity in STEM and to develop a road map for how schools, organizations outside of schools, and universities can challenge talented students during their scientifically formative years—adolescence and early adulthood—and recommend a research program to rigorously study their effectiveness.

Topics for Study

An expert panel discussion would involve a range of goals, such as:

- Identifying strategies for nurturing the talents of those individuals in adolescence and early adulthood who are likely to become the next generation of high-level STEM professionals and innovators.
- Exploring the possible existence of pools of potential talent in our society that currently are overlooked, under-developed, and under-utilized, but who could become a source of adults productive in STEM and who could fuel innovation in this country.
- Creating a research agenda on effective means for nurturing and developing the STEM talent in youth and early adulthood in order to accelerate the STEM productivity and creativity of such individuals over their careers.
- Suggesting and encouraging development of policies that could help ensure a strong pipeline of STEM talent and nurture innovation in the STEM workforce.

Logistics

The National Science Board Office will be the focal point for providing all aspects of Board support for this Board activity; coordinating NSF, the involvement of other agencies and institutions; and utilizing contractual or NSB Office staff resources to support events in connection with this Board- sponsored activity.

An agenda and a comprehensive list of potential participants for the two-day expert panel discussion will be developed with input from Board Members, NSF management, and other knowledgeable sources in the broader STEM research and education community.

CEH leadership

NSB/CEH will recommend full Board approval of the appointment of an *ad hoc* Task Group of CEH to provide oversight for, and actively engage in, this activity, on behalf of the CEH Committee with membership including: Drs. Camilla Benbow, John Bruer, José-Marie Griffiths, Louis Lanzerotti, and Diane Souvaine.

Product

The final output from this activity will be a concise set of Board-approved recommendations to NSF (and perhaps to other Federal entities), informed by a white paper capturing the results of the expert panel discussion and reflecting input from NSF and other agency expert staff, written background materials addressing these issues, and comments from interested communities on initial, Board- approved draft recommendations.

Schedule

A final, concise report will be submitted to the Board for approval and publication by summer 2010.

Audience

In addition to the President, Congress, and NSF:

- Federal agencies involved in STEM education
- State and local organizations and individuals involved or interested in STEM education
- Educational and professional organizations with interests in STEM education
- Employers of STEM-educated workers

APPENDIX II

Expert Panel Discussion Participants (August 23-25, 2009)

The Honorable Arne Duncan, U.S. Secretary of Education (Keynote Address)

National Science Board Members (in alphabetical order)

Dr. Arden L. Bement, Jr., Director, National Science Foundation and NSB Member *ex officio*

Dr. Camilla P. Benbow, Member; Patricia and Rodes Hart Dean of Education and Human Development, Peabody College, Vanderbilt University

Dr. John T. Bruer, Member & Chairman of the Committee on Education and Human Resources; President, The James S. McDonnell Foundation

Dr. Patricia D. Galloway, Vice Chairman, National Science Board; Chief Executive Officer, Pegasus Global Holdings, Inc., Cle Elum, Washington

Dr. José-Marie Griffiths, Member; Dean and Professor, School of Information and Library Science; Director of Biomedical Informatics, TraCS Institute, School of Medicine, University of North Carolina, Chapel Hill

Dr. Louis J. Lanzerotti, Member; Distinguished Research Professor of Physics, Center for Solar Terrestrial Research, Department of Physics, New Jersey Institute of Technology

Dr. Douglas D. Randall, Member; Professor and Thomas Jefferson Fellow, University of Missouri, Columbia

Dr. Diane L. Souvaine, Member; Department Chair and Professor of Computer Science, Tufts University

Dr. Kathryn D. Sullivan, Member; Director, Battelle Center for Mathematics and Science Education Policy, John Glenn School of Public Affairs, Ohio State University, Columbus

Dr. Craig R. Robinson, Acting Executive Officer, National Science Board and National Science Board Office Director

Panelists/discussants (in alphabetical order)

Dr. Michael J. Cima, Professor of Engineering, Massachusetts Institute of Technology; Director, Lemelson-MIT Invention and Innovation Center

Dr. Nicholas Colangelo, Director, The Connie Belin & Jacqueline N. Blank International Center for Gifted Education and Talent Development, University of Iowa

Dr. Diane C. DiEuliis, Assistant Director, Life Sciences, Office of Science and Technology Policy

Ms. Patricia Johnson, Javits Gifted and Talented Students Education Program, U.S. Department of Education

Mr. Dean Kamen, President, DEKA Research & Development

Dr. Ken Kotovsky, Professor and Director of Undergraduate Studies in Psychology, Carnegie Mellon University

Dr. David F. Lohman, Professor of Educational Psychology, University of Iowa

Dr. David Lubinski, Professor of Psychology, Peabody College; Co-Director, Study of Mathematically Precocious Youth

Dr. Cora B. Marrett, Deputy Director (acting), National Science Foundation

Dr. Stephanie Pace Marshall, Founding President of the Illinois Mathematics and Science Academy and founding President of the National Consortium for Specialized Secondary Schools of Mathematics, Science and Technology

Ms. Zipporah A. Miller, Associate Executive Director, Professional Programs and Conferences, National Science Teachers Association

Dr. Arthur P. Molella, Director, Lemelson Center for Invention and Innovation, National Museum of American History, Smithsonian

Dr. Diana G. Oblinger, President and Chief Executive Officer, Educause

Dr. Diana Rhoten, Program Director, Knowledge Institutions; Research Director, Digital Media and Learning, Social Science Research Council

Dr. Ann Robinson, Professor of Education and founding Director of the Center for Gifted Education, University of Arkansas at Little Rock; 2008-2009 President, National Association for Gifted Children

Dr. Robert Root-Bernstein, Professor of Physiology, Michigan State University

Dr. R. Keith Sawyer, Associate Professor, Department of Education, Washington University

Dr. Larisa V. Shavinina, Professor of Project Management & Innovation, Department of Administrative Sciences, Université du Québec en Outaouais (UQO), Canada

Dr. Sally Goetz Shuler, Executive Director, National Science Resource Center

Dr. Rena F. Subotnik, Director, Center for Psychology in Schools and Education, American Psychological Association

Dr. Joyce VanTassel-Baska, Director, Center for Gifted Education, The College of William and Mary

Dr. Jo Anne Vasquez, Vice President and Program Director Teacher & Curriculum Initiatives, Helios Education Foundation

Dr. Frank C. Worrell, Professor, University of California Berkeley; Faculty Director, Academic Talent Development Program; Director of Research and Development, California College Preparatory Academy

Mr. Joshua Wyner, Senior Vice President (Policy), National Consortium for College Completion

Dr. Lea Ybarra, Executive Director, Center for Talented Youth, Johns Hopkins University

Student Lunch Panel (in alphabetical order)

Richard Li, River Hill High School, Howard County, Maryland, Class of 2010

Elena Perry, Richard Montgomery High School, Montgomery County, Maryland, Class of 2010

Andrew Das Sarma, Montgomery Blair High School, Montgomery County, Maryland, Class of 2011

Louis Wasserman, University of Chicago, Class of 2012

Dr. Alex Wissner-Gross, Environmental Fellow, Harvard University

Additional online resources relating to the August 23-25, 2009 expert panel discussion can be found at the following URL: http://www.nsf.gov/nsb/meetings/2009/0824/index.jsp

APPENDIX III

NSB/CEH-09-06
August 25, 2009

EXPERT PANEL DISCUSSION ON PREPARING THE NEXT GENERATION OF STEM INNOVATORS

FINAL AGENDA
Sunday, August 23

6:00 – 8:30 **Welcome Discussion & Dinner**
Dan & Brad's restaurant, Arlington VA

~6:45-7:15 **Overview: What is the State of the Field?**
Summary of gifted and talented education, innovation, creative thinking, learning sciences and the current state of our educational system with regard to these topics
Presenter: **Dr. Joyce VanTassel-Baska**, Director, Center for Gifted Education, The College of William and Mary

7:15-7:30 *Reaction:* **Dr. Rena Subotnik**, Director, Center for Psychology in Schools and Education, American Psychological Association

7:30-8:30 Discussion

Monday, August 24

8:00 **Welcome**
Dr. Patricia D. Galloway, Vice Chairman, National Science Board
Dr. John T. Bruer, Chairman, Committee on Education and Human Resources (CEH), STEM Innovators Task Group, National Science Board
Dr. Camilla P. Benbow, Lead, STEM Innovators Task Group, Committee on Education and Human Resources, National Science Board

8:15 **Board Process and Participant Introductions**

8:20 – 3:00 **Session I: Characterization and Development of Future STEM Innovators**

8:20-9:50 **Cognitive and non-cognitive characteristics of an innovator**
Guiding questions: What are some of the defining characteristics of an innovator and potential future innovators? How important are attributes such as ability, interest, determination, and inquisitiveness? How can theories of cognition, motivation, and other non-cognitive factors be applied to educational practices for fostering innovation? What do research on inquiry in science education and theories of intelligence and innovation add to the discussion? What research needs to be done to determine the most effective means (both cognitive and non-cognitive) for identifying STEM talent in youth and early adulthood? What are the implications for policy?
Moderator: **Dr. Camilla P. Benbow**

Panelists:

- **Dr. David Lubinski**, Professor of Psychology, Peabody College; Co-Director, Study of Mathematically Precocious Youth
- **Dr. R. Keith Sawyer**, Associate Professor, Department of Education, Washington University
- **Dr. Larisa V. Shavinina**, Professor of Project Management & Innovation at the Department of Administrative Sciences, Université du Québec en Outaouais (UQO), Canada

9:50 **Break**

10:00 -12:00 Developing STEM innovators through the education system

Guiding questions: Once we understand the characteristics of a potential innovator, how do we 1) initiate the innovation process and 2) develop a possible STEM innovator in order to increase the likelihood of productivity over an entire career? What kinds of schools or *formal* learning settings are best for motivating students to become STEM innovators? How can we expand the kinds of opportunities that have promising evidence on effectiveness to broader populations of students? How do we raise the ceiling of potential for the exceptionally gifted and/or motivated student? How can we best anticipate future learning environments? Why do talent losses occur at critical transition points in the educational system? How can higher education best partner with other institutional components of the innovation life cycle? What are the policy implications?

Moderator: **Dr. Diane L. Souvaine**

Panelists:

- **Dr. Nicholas Colangelo**, Director, The Connie Belin & Jacqueline N. Blank International Center for Gifted Education and Talent Development, University of Iowa
- **Dr. Stephanie Pace Marshall**, founding President of the Illinois Mathematics and Science Academy and founding President of the National Consortium for Specialized Secondary Schools of Mathematics, Science and Technology
- **Dr. Robert Root-Bernstein**, Professor of Physiology, Michigan State University
- **Dr. Lea Ybarra**, Executive Director, Center for Talented Youth, Johns Hopkins University

12:00 **Lunch**: Perspective from current and former students

Lunch will be provided to invited panelists and discussants only

Guiding questions: What has been your experience in the education system? Do/did you feel sufficiently challenged? Are you aware of and encouraged by your school to take advantage of enrichment opportunities, such as laboratory research partnerships, summer programs, or other opportunities such as accelerated learning? What was the most important factor in seeding your interest in the STEM disciplines? What was the biggest challenge you faced or what was the most significant negative force in terms of your education? How would you change it? What helped transform your creative potential into reality? What hinders it?

Moderator: **Dr. Kathryn D. Sullivan**

Panelists:

Introduction: **Dr. Carol Blackburn**, Johns Hopkins University

- **Richard Li**, River Hill High School, Class of 2010
- **Elena Perry**, Richard Montgomery High School, Class of 2010
- **Andrew Das Sarma**, Montgomery Blair High School, Montgomery County, Maryland, Class of 2011
- **Louis Wasserman**, University of Chicago, Class of 2012
- **Dr. Alex Wissner-Gross**, Environmental Fellow, Harvard University

1:15-3:15 Informal learning, cyber-learning and innovative education

Guiding questions: What kinds of *informal* learning settings are effective for motivating students to develop the skills needed to become a potential STEM innovator? How can we expand the kinds of opportunities that have promising evidence on effectiveness to more and broader populations of high-potential students? How can new technologies be harnessed to serve the development and possibly enhance productivity of future STEM innovators? How can these emerging technologies be used to foster collaboration, enhance networking across multiple disciplines, and generate improvements in both informal and traditional learning environments that might nurture STEM innovation potential? What are the policy implications?

Moderator: **Dr. José-Marie Griffiths**

Panelists:

- **Dr. Arthur P. Molella**, Director, Lemelson Center for Invention and Innovation, National Museum of American History, Smithsonian
- **Dr. Diana G. Oblinger**, President and CEO of EDUCAUSE
- **Dr. Diana Rhoten**, Program Director, Knowledge Institutions; Research Director, Digital Media and Learning, Social Science Research Council

3:15 Break

3:30 – 5:30 Session II: Identifying and Nurturing Under-developed STEM Talent

Identifying under-developed pools of STEM talent and the community role in fostering achievement

Guiding questions: How can we best identify and nurture pools of potential STEM talent in our society that currently are overlooked, under-developed, and under-utilized, but could become a source of adults productive in STEM and could fuel innovation in this country? What role does the community (parents, teachers, local businesses) play in nurturing, supporting and motivating students? Do ethnically or geographically distinct subgroups of students learn differently? What role does cultural background play in talent development? What are the policy implications?

Moderator: **Dr. Louis J. Lanzerotti**

Panelists:

- **Dr. Rena F. Subotnik**, Director, Center for Psychology in Schools and Education, American Psychological Association
- **Dr. David F. Lohman**, Professor of Educational Psychology, University of Iowa
- **Dr. Frank C. Worrell**, Professor, UC Berkeley. Faculty Director, Academic Talent Development Program, Director of Research and Development, California College Preparatory Academy
- **Mr. Joshua Wyner**, Senior Vice President (Policy), National Consortium for College Completion

5:30 **Dinner on your own**

Tuesday, August 25

8:00 – 8:30 Keynote Address

Introduction: Dr. Arden L. Bement, Jr., Director, NSF

Keynote Address: The Honorable Arne Duncan, U.S. Secretary of Education

8:30 – 10:30 Session III: The Products of Innovation

The innovation ecology and entrepreneurship

Guiding questions: There are many factors external to the individual involved with innovation. Innovations do not occur in a vacuum, and an innovation can change its own environment. What does the research say about innovation as a product of individuals or a product of groups? How would the collaborative process factor in the learning processes associated with innovation? What can we learn from industry and business regarding innovation and entrepreneurship that would be helpful in improving formal and informal learning environments? How can lessons learned from these groups improve policy-making?

Moderator: **Dr. John T. Bruer**

Panelists:

- **Dr. Michael J. Cima**, Sumitomo Electric Industries Professor of Engineering, MIT; Director, Lemelson-MIT Invention and Innovation Center
- **Mr. Dean Kamen**, President, Deka Research & Development
- **Dr. Kenneth Kotovsky**, Professor and Director of Undergraduate Studies in Psychology, Carnegie Mellon University

10:30 **Break**

10:45–12:45 Session IV: Perspectives on Government Education Programs and Policy

Existing government education programs, program assessment and effective policy design and implementation

Guiding questions: What are we currently doing for innovators in terms of programs and policies? What types of policy recommendations are ideal in terms of implementation? Can state/local government education policy inform Federal Government policy recommendations? How do we define success in STEM education (e.g., PISA scores, other metrics) and how does STEM education success in the U.S. compare internationally? What can we learn from successful international STEM education systems, particularly with regard to high-ability students?

Moderator: **Dr. Camilla P. Benbow**

Panelists:

- **Dr. Diane C. DiEuliis**, Assistant Director, Life Sciences, OSTP
- **Ms. Patricia Johnson,** U.S. Department of Education, Javits Gifted and Talented Students Education Program
- **Dr. Cora B. Marrett,** Deputy Director (acting), NSF

12:45 Adjourn

Additional Discussion Participants

- **Ms. Zipporah A. Miller**, Associate Executive Director, Professional Programs and Conferences, National Science Teachers Association
- **Dr. Ann Robinson**, Professor of Education, founding Director of the Center for Gifted Education, University of Arkansas at Little Rock; 2008-2009 President, National Association for Gifted Children
- **Dr. Sally Goetz Shuler**, Executive Director, National Science Resource Center
- **Dr. Jo Anne Vasquez**, Vice President and Program Director Teacher & Curriculum Initiatives, Helios Education Foundation

COVER IMAGE CREDITS

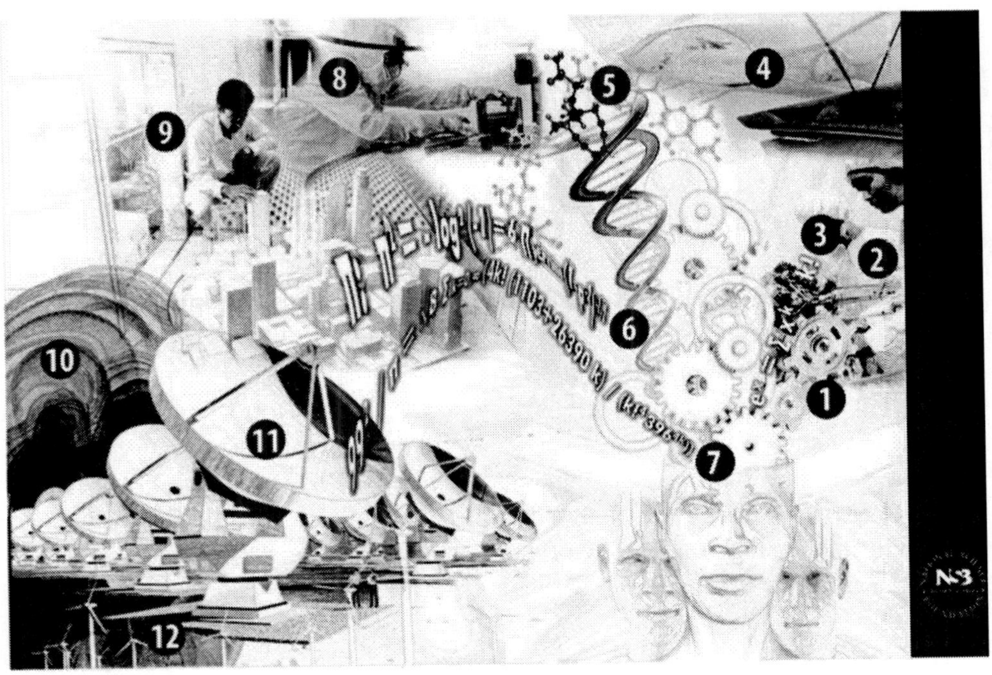

1) These miniature gears, developed by researchers at AT&T Bell Laboratories, are about the size of a human hair and are driven by air forced through their ports. Micromechanics is one of the emerging technologies supported by the National Science Foundation (NSF).
   Credit: AT&T Bell Labs
2) A West Virginia University electrical engineering graduate research assistant explains the use of molecular beam epitaxy for the growth of nanostructures to a physics undergraduate. The equipment was upgraded using funding from the NSF EPSCoR (Experimental Program to Simulate Competitive Research) activity, provided through West Virginia EPSCoR during a previous Research Infrastructure Improvement grant. The research being performed was funded by NSF.
   Credit: West Virginia University - WVNano (Date of image: 2004)
3) Sir Isaac Newton (1643-1727; England) made revolutionary advances in mathematics, optics, dynamics, thermodynamics, acoustics and celestial mechanics. In addition to several other important advances in analytic geometry, his mathematical works include the Binomial Theorem, his eponymous numeric method, the idea of polar coordinates, and power series for exponential and trigonometric functions. His equation $ex = \sum xk / k!$ has been called the most important series in mathematics.
4) A composite of scanning electron microscope images showing biological force microscopy, developed by NSF- funded researchers at Virginia Polytechnic Institute and Virginia State University.
   Credit: Dr. Steven Lower, University of Maryland; and Dr. Michael Hochella, Virginia Tech
5) A molecular model of esperamicin A1, an enediyne. Enediynes are naturally occurring molecules—commonly called biological warheads—for their ability to bind to and split tumor's DNA backbones. Computations were performed on the National Center for Supercomputing Applications (NCSA) SGI Origin2000 supercomputer, purchased primarily with funds from NSF.
   Credit: Images by Steven Feldgus; simulation completed using computational resources provided by the NCSA [Structure comes from Kumar, R. A., Ikemoto, N., & Portel, D. J. (1997). *J. Mol. Biol, 265*, 173-186.]
6) Leonhard Euler (1707-1783; Switzerland) made decisive contributions in all areas of mathematics; he gave the world modern trigonometry. Along with Lagrange, he pioneered the calculus of variations. He was the most prolific mathematician in history and the best algorist. Some of Euler's greatest formulae can be combined into curious- looking formulae for π: $\pi 2 = -\log 2(-1) = 6 \prod p \in Prime(1-p-2)-1/2$.
7) Srinivasa Ramanujan Iyengar (1887-1920; India) was a self-taught prodigy who lived in a country distant from his mathematical peers, and suffered from poverty and malnutrition. Despite these limitations, Ramanujan is considered one of the greatest geniuses ever and produced 4000 theorems or conjectures in number theory, algebra, and combinatorics. Because of its fast convergence, an odd-looking formula of Ramanujan is often used to calculate π: $99^2 / \pi = \sqrt{8} \sum k=0,\infty (4k! (1103+26390 k) / (k!4\ 3964k))$.
8) A technician at Texas Instruments processes wafers containing computer microchips. The Maricopa Advanced Technology Education Center (MATEC) effects change in

technician education through the creation of competency-based curricula, diverse and effective professional development programs, and replicable workforce development models. MATEC is funded in part by a grant from NSF's Advanced Technological Education (ATE) program.
Credit: Photo from ATE Centers Impact 2008-2010, *http://www.atecenters.org* (Date of image: 2009)

9) A model of downtown Minneapolis, Minnesota, is prepared for testing in the boundary layer wind tunnel at the University of Minnesota's St. Anthony Falls Laboratory (SAFL). The test will study the effects of wind on structures, including stress on windows, heat loss due to leaks and poor air conditioning and ventilation system performance, as well as the effects of wind on pedestrians. Support for the design and construction of the wind tunnel was made through an NSF grant.
Credit: Courtesy Pat Swanson, St. Anthony Falls Laboratory, University of Minnesota
10) This numerical simulation is part of a series depicting orbiting black holes and represents the first time that three- quarters of a full orbit has been computed. The simulations show the merger of two black holes and the ripples in space-time—known as gravitational waves—that are born of the merger. These simulations were created on the NCSA Itanium Linux Cluster (It) by researchers from the Max Planck Institute for Gravitational Physics (Albert Einstein Institute) in Potsdam, Germany, and visualized by Werner Benger of the Albert Einstein Institute (AEI) and the Konrad-Zuse-Zentrum in Berlin. NSF support was used for this project.
Credit: Simulations by Max Planck Institute for Gravitational Physics (Albert-Einstein-AEI); visualization by Werner Benger, Zuse Institute, Berlin and AEI. The computations were performed on NCSA's It.
11) The image represents an artist's conception of the antennas for the Atacama Large Millimeter Array (ALMA). The construction and operation of ALMA will be funded through a joint agreement between NSF and the European Southern Observatory. NSF will execute the project through the National Radio Astronomy Observatory (NRAO). ALMA will be an array of 64, 12-meter radio antennas that will work together as one telescope to study millimeter and sub-millimeter wavelength light from space. These wavelengths of the electromagnetic spectrum, which cross the critical boundary between infrared and microwave radiation, hold the key to understanding such processes as planet and star formation, the formation of early galaxies and galaxy clusters and the detection of organic and other molecules in space.
12) This image of wind turbines contrasts the past—a wind pump used to draw water from farm wells for cattle, and the future—modern wind turbines. The wind turbines are part of the Cedar Creek wind farm in Colorado. The farm includes more than 250 turbines and generates roughly 300 megawatts of energy. As wind energy grows in importance, scientists at the National Center for Atmospheric Research (NCAR) are studying how wind turbines and farms interact with the atmosphere, and how their output can be better predicted and managed.
Credit: University Corporation for Atmospheric Research (Date of image: unknown)

NATIONAL SCIENCE BOARD RECENT PUBLICATIONS

Science and Engineering Indicators 2010 (volume 1, NSB-1 0-01; volume 2, NSB- 10-0 1A)

Building a Sustainable Energy Future: U.S. Actions for an Effective Energy Economy Transformation (NSB-09-55)

Investing in the Future: NSF Cost Sharing Policies for a Robust Federal Research Enterprise (NSB-09-20)

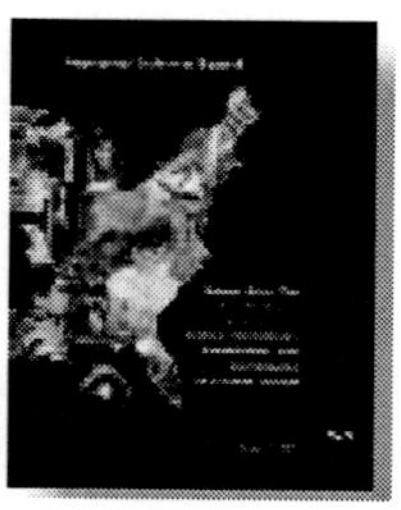

A National Action Plan for Addressing the Critical Needs of the U.S. Science, Technology, Engineering, and Mathematics Education System (NSB-07-114)

End Notes

[1] Roosevelt, F. D. (1945). President Roosevelt's letter. In V. Bush, *Science–the endless frontier. A report to the President on a program for postwar scientific research* (p. 4). Washington, DC: U.S. Government Printing Office.

[2] Bush, V. (1945). *Science–the endless frontier. A report to the President on a program for postwar scientific research* (p. 23). Washington, DC: U.S. Government Printing Office.

[3] The National Commission on Excellence in Education. (1983). *A nation at risk: The imperative for educational reform.* Washington, DC: U.S. Department of Education. Retrieved from: http://www2.ed.gov /pubs/NatAtRisk/index.html.

[4] National Academy of Sciences. (2005). *Rising above the gathering storm: Energizing and employing America for a brighter economic future*. Washington, DC: The National Academies Press.

[5] Ibid. See Chapter 7: "What actions should America take in science and engineering higher education to remain prosperous in the 21st century?" pp. 162-181. Also, see the issue brief, "Attracting the most able US students to science and engineering," pp. 325-341.

[6] The *STEM Innovators* report is not primarily focused on what is classically thought of as "gifted and talented" (G&T), though the G&T scholarly community and research compendium have informed the present project. The Board recognizes that within the G&T scholarly community, the terms "gifts" or "gifted" and "talents" are not used interchangeably. However, for simplicity, the Board has elected to group these terms together.

[7] Lohman, D. F. (1994). Spatial ability. In R. J. Sternberg (Ed.), *Encyclopedia of intelligence* (p. 1000). New York, NY: Macmillan Press. In this chapter, the author defines spatial ability as, "the ability to generate, retain, retrieve, and transform well-structured visual images." Note: In the present report, spatial ability includes mechanical reasoning.

[8] Bush, V. (1945). *Science–the endless frontier. A report to the President on a program for postwar scientific research* (p. 18). Washington, DC: U.S. Government Printing Office.

[9] National Science Board. (2010). Industry, technology, and the global marketplace. In *Science and engineering indicators 2010*. Arlington, VA: National Science Foundation. Retrieved from: http://www.nsf.gov/statistics

[10] Executive Office of the President. (2010). *A strategy for American innovation: Driving towards sustainable growth and quality jobs* (p. 4). Washington, DC: Office of Science and Technology Policy, National Economic Council. Retrieved from: http://www.whitehouse.gov/sites/default/files/microsites/20090920-innovation-whitepaper.pdf.

[11] Council on Competitiveness. (2007). *Competitive index: Where America stands* (p. 10). Washington, DC: Author. Retrieved from:
http://www.compete.org/images March_2007.pdf.

[12] Lubinski, D., & Benbow, C. P. (2006). Study of mathematically precocious youth after 35 years: uncovering antecedents for the development of math-science expertise. *Perspectives on Psychological Science*, *1*, 316-345.

[13] Baldi, S., Jin, Y., Skemer, M., Green, P. J., & Herget, D. (2007). *Highlights from PISA 2006: Performance of U.S. 15-year-old students in science and mathematics literacy in an international context* (NCES 2008–0 16) (pp. 44, 49). National Center for Education Statistics, Institute of Education Sciences. Washington, DC: U.S. Department of Education. Retrieved from: http://nces.ed.gov/pubs2008/2008016.pdf. Note: The rank number represents the combination of OECD and non-OECD jurisdictions.

[14] Gonzales, P., Williams, T., Jocelyn, L., Roey, S., Kastberg, D., & Brenwald, S. (2008). *Highlights from TIMSS 2007: Mathematics and science achievement of U.S. fourth- and eighth-grade students in an international context* (NCES 2009-00 1 Revised) (p. 16). National Center for Education Statistics, Institute of Education Sciences. Washington, DC: U.S. Department of Education. Retrieved from: http://nces.ed.gov/pubs2009/2009001.pdf.

[15] Ibid.

[16] Zumeta, W., & Raveling, J. (2002). *The best and brightest: Is there a problem here?* (pp. 4-6). Washington, DC: Commission on Professionals in Science and Technology. Retrieved from: *http://www.cpst.org/BBIssues.pdf*. In this analysis, "highest achievers" refers to U.S. citizens scoring over a 750 on the GRE quantitative scale. The authors caution that these data are limited because "they are based on individuals' responses on the GRE registration questionnaire as to their intended field of graduate study, not on their actual enrollment behavior." (See p. 4).

[17] Lowell, B. L., Salzman, H., Bernstein, H. H., & Henderson, E. (2009). *Steady as she goes? Three generations of students through the science and engineering pipeline*. Paper presented at Annual Meetings of the Association for Public Policy Analysis and Management Washington, DC on November 7, 2009. Retrieved from: http://www.heldrich.rutgers. edu/uploadedFiles/Publications/STEM_Paper_Final.pdf. In this report, for the high school-to-college transition point, "highest performers" refers to high school students scoring in the top quintile of the SAT or ACT (pp. 9-10). The data showing a decline in the numbers of highest-achieving students enrolling in a STEM major or completing college with a STEM degree are shown in Figure 1 (p. 17). According to the data, there was a 14.9 percent decline between the 1992/97 cohort and the 2000/05 cohort. For further discussion regarding the high school-to-college transition point, see pp. 16-20.

[18] National Science Board. (2003). *The science and engineering workforce – realizing America's potential* (NSB-03-69). Arlington, VA: National Science Foundation. Retrieved from: http://www.nsf.gov/nsb/documents/2003/nsb0369/nsb0369.pdf.

[19] National Science Board. (2010). *Science and engineering indicators 2010: Appendix tables* (pp. 128-130). Arlington, VA: National Science Foundation. Retrieved from: http://www.nsf.gov/statistics. See Table 2-35, "First university degrees, by selected region and country/economy: 2006 or most recent year."

[20] National Science Board. (2010). Higher education in science and engineering. In *Science and engineering indicators 2010* (chapter 2, pp. 25, 31). Arlington, VA: National Science Foundation. Retrieved from: http://www.nsf.gov/statistics

[21] National Science Board. (2010). Science and engineering labor force. In *Science and engineering indicators 2010* (chapter 3, p. 52). Arlington, VA: National Science Foundation. Note: Foreign-born workforce data for 2003 are located in Table 3-24. Retrieved from: http://www.nsf.gov/statistics

[22] Wai, J., Lubinski, D., & Benbow, C. P. (2009). Spatial ability for STEM domains: Aligning over 50 years of cumulative psychological knowledge solidifies its importance. *Journal of Educational Psychology, 101*, 8 17-835.

[23] Ibid.

[24] For a discussion on college-age demographics, see: National Science Board. (2010). Higher education in science and engineering. In *Science and engineering indicators 2010* (chapter 2, p. 13). Arlington, VA: National Science Foundation. Retrieved from: http://www.nsf.gov/statistics. Also, see: Hussar, W. J., & Bailey, T. M. (2008). *Projections of education statistics to 2017* (NCES 2008-078). National Center for Education Statistics, Institute of Education Sciences. Washington, DC: U.S. Department of Education. Retrieved from: http://nces.ed.gov/pubs2008/2008078.pdf.

[25] National Science Board. (2010). Higher education in science and engineering. In *Science and engineering indicators 2010* (chapter 2, p. 24). Arlington, VA: National Science Foundation. Retrieved from: http://www.nsf.gov/statistics seind10/pdf/seind10.pdf. Also, see Appendix Table 2-28.

[26] Business Roundtable. (2008). *Tapping America's potential: The education for innovation initiative* (p. 5). Washington, DC: Author. Retrieved from: http://www.eric.ed.gov/ERICDocs/data/ericdocs2sql/content_storage ad/71.pdf.

[27] Bush, V. (1945). *Science–the endless frontier. A report to the President on a program for postwar scientific research.* Washington, DC: U.S. Government Printing Office. Quote originates from the Vannevar Bush Committee on Discovery and Development of Scientific Talent, p. 25.

[28] For example, seven Intel Science Talent Search finalists have been awarded the Nobel Prize, two have earned the Fields Medal, and 30 have been elected to the National Academy of Sciences. For more awards and honors of Intel Science Talent Search finalists, see: Society for Science & The Public. *Alumni honors.* http://www.societyforscience.org/sts/alumni.

[29] Wai, J., Lubinski, D., Benbow, C. P., & Steiger, J. H. (2010). Accomplishment in science, technology, engineering, and mathematics (STEM) and its relation to STEM educational dose: A 25-year longitudinal study. *Journal of Educational Psychology, 102* [In Press].

[30] The Council of the State Directors of Programs for the Gifted & The National Association for Gifted Children. (2010). *State of the states in gifted education: National policy and practice data 2008-2009*. Washington, DC: Author. Methodology note: 47 out of 50 states responded to the 2008-2009 questionnaire. Of those, the responses of 45 states were considered complete.

[31] U.S. Department of Education. *Fiscal Year 2010 Budget Summary – May 7, 2009.* http://www.ed.gov/about/overview/ budget/budget10/summary/edlite-section4.html.

[32] Some NSF programs, such as the *Graduate Research Fellowships Program* and *NSF Scholarships in Science, Technology, Engineering, and Mathematics*, provide support for talented individuals at the undergraduate, graduate and post-doctoral levels.

[33] Reis, S. M., Hébert, T. P., Díaz, E. I., Maxfield, L. R., & Ratley, M. E. (1995). *Case studies of talented students who achieve and underachieve in an urban high school* (RM95 120). Storrs, CT: The National Research Center on the Gifted and Talented, University of Connecticut. Retrieved from: http://www.gifted

[34] Wyner, J. S., Bridgeland, J. M., & DiIulio, J. J. (2007). *Achievement trap: How America is failing millions of high- achieving students from lower-income families*. Landsdowne, VA: Jack Kent Cooke Foundation. Retrieved from: http:// www.civicenterprises.net/pdfs/jkc.pdf.

[35] Loveless, T. (2008). Analysis of NAEP data. In A. Duffet, S. Farkas, & T. Loveless (Eds.), *High-achieving students in the era of NCLB* (pp. 13-48). Washington, DC: Thomas B. Fordham Institute. Retrieved from: http://www.edexcellence.net/doc/20080618_high_achievers_part1.pdf.

[36] Wyner, J. S., Bridgeland, J. M., & DiIulio, J. J. (2007). *Achievement trap: How America is failing millions of high- achieving students from lower-income families* (p. 11). Landsdowne, VA: Jack Kent Cooke Foundation. Retrieved from: http://www.civicenterprises.net/pdfs/jkc.pdf.

[37] Ibid. For a discussion of the authors' methodology, refer to p. 8. For a graphical representation of the educational disparities among high achievers, see p. 6.

[38] Ibid., p. 12.

[39] Plucker, J. A., Burroughs, N., & Song, R. (2010). *Mind the (other) gap! The growing excellence gap in K-12 education*. Center for Evaluation and Education Policy. Bloomington, IN: Indiana University School of Education. Retrieved from: https://www.iub.edu/~ceep/Gap/excellence/ExcellenceGapBrief.pdf. This report examines the achievement gap—or what the authors call the "excellence gap"—based on gender, socioeconomic status, race/ethnicity, and English language proficiency. For data analysis purposes, whether the achievement gap at the high end is widening, stable, or narrowing depends on how the top cohort is defined. For example, gaps among those reaching the "advanced" level in mathematics on the NAEP are generally widening, while gaps among those reaching the 90th percentile in mathematics on the NAEP are generally stable and in some cases narrowing. However, even if these gaps are narrowing, the authors point

out that the rate of advancement of the underserved groups is so slow that it would require several decades to close them. See pp. 4-13 for a discussion on NAEP national data, and pp. 13-15 for NAEP state data. See pp. 15-18 for a discussion on the use of proficiency level cut-points (e.g., below basic, basic, proficient, advanced) compared to the use of percentiles (e.g., 90th percentile).

[40] Ibid., p. 6.

[41] See the following four references for an examination of the issue of educational achievement gaps: 1) Reardon, S. (2008). *Differential growth in the black-white achievement gap during elementary school among high- and low-scoring students* (Working Paper 2008-7). Stanford, CA: Institute for Research on Education Policy & Practice, Stanford University. Retrieved from: http://www.stanford.edu/group/irepp/cgi-bin/joomla/docman/differential-growth-in-the-blackwhite-achievement-gap-reardon/download. html. 2) Hanushek, E. A., & Rivkin, S. G. (2009). Harming the best: How schools affect the black-white achievement gap. *Journal of Policy Analysis and Management*, *28*, 366-393. 3) Donovan, M. S., & Cross, C. T. (2002). *Minority students in special and gifted education*. National Research Council. Committee on Minority Representation in Special Education. Division of Behavioral and Social Sciences and Education. Washington, DC: National Academy Press. 4) Gándara, P. (2005). *Fragile futures: Risk and vulnerability among Latino high-achievers*. Princeton, NJ: Policy Evaluation and Research Center, Educational Testing Service. Retrieved from: http://www.ets.org/Media/Research/pdf/PICFRAGFUT.pdf.

[42] See the following three references regarding the underrepresentation of African American, Hispanic, and Indian/Native Alaskan students in gifted classrooms, AP classes, and achievement gaps at the "high end.": 1) Gándara, P. (2005). *Fragile futures: Risk and vulnerability among Latino high-achievers*. Princeton, NJ: Policy Evaluation and Research Center, Educational Testing Service. Retrieved from: http://www.ets.org/Media/Research/pdf/PICFRAGFUT.pdf. 2) Learning Point Associates. (2004). *All students reaching the top: Strategies for closing academic achievement gaps*. Naperville, IL: North Central Regional Educational Laboratory. Retrieved from: http://www.ncrel.org/gap/studies/allstudents.pdf. 3) Hoffman, K., & Llagas, C. (2003). *Status and trends in the education of blacks* (NCES 2003–034). Project Officer: T. D. Snyder. National Center for Education Statistics. Washington, DC: U.S. Department of Education. Retrieved from: http://nces.ed.gov/pubs2003/2003034.pdf.

[43] The White House, Office of the Press Secretary. (2009). *Remarks by the President on the 'educate to innovate' campaign*. Speech given by U.S. President, Barack Obama, on November 23, 2009. Retrieved from: http://www.whitehouse.gov/the-press-office/remarks-president

[44] Benbow, C. P., & Stanley, J. C. (1996). Inequity in equity: How "equity" can lead to inequity for high-potential students. *Psychology, Public Policy, and Law*, *2*, 249-292.

[45] Farkas, S., & Duffet, A. (2008). Results from a national teacher survey. In S. Farkas, A. Duffet, & T. Loveless (Eds.), *High-achieving students in the era of NCLB* (pp. 49-82). Washington, DC: Thomas B. Fordham Institute. Retrieved from: http://www.edexcellence.net/doc/20080618_high_achievers_part2.pdf.

[46] For a discussion on teacher training and talented students, see: The Council of the State Directors of Programs for the Gifted & The National Association for Gifted Children. (2010). *State of the states in gifted education: National policy and practice data 2008-2009* (pp. 38-41). Washington, DC: Author. Also, see: Archambault, F. X., Jr., Westberg, K. L., Brown, S., Hallmark, B. W., Emmons, C., & Zhang, W. (1993). *Regular classroom practices with gifted students: Results of a national survey of classroom teachers* (RM93 102). Storrs, CT: The National Research Center on the Gifted and Talented, University of Connecticut. Retrieved from: http://www.gifted.uconn.edu/nrcgt/reports/rm93102/rm93102.pdf.

[47] Archambault, F. X., Jr., Westberg, K. L., Brown, S., Hallmark, B. W., Emmons, C., & Zhang, W. (1993). *Regular classroom practices with gifted students: Results of a national survey of classroom teachers* (RM93 102). Storrs, CT: The National Research Center on the Gifted and Talented, University of Connecticut. Retrieved from: http://www.gifted.uconn.edu/nrcgt/reports/rm93102/rm93102.pdf. Sixty-one percent of approximately 7300 randomly selected third and fourth grade teachers in public and private schools in the United States reported that they had never had any training in teaching gifted students.

[48] National Science Board. (2007). *Moving forward to improve engineering education* (NSB-07-1 22). Arlington, VA: National Science Foundation. Retrieved from: http://www.nsf.gov/pubs/2007/nsb07122/nsb07122.pdf.

[49] Marra, R. M., Rodgers, K. A., Shen, D., & Bogue, B. (2009). Women engineering students and self-efficacy: a multiyear, multi-institution study of women engineering student self-efficacy. *Journal of Engineering Education*, *98*(1), 1-12.

[50] Ford D. Y., Grantham, T. C., & Whiting, G. W. (2008). Culturally and linguistically diverse students in gifted education: recruitment and retention issues. *Exceptional Children*, *74*, 289-306.

[51] Ibid.

[52] Dweck, C. S. (2008). *Mindsets and math/science achievement*. Paper prepared for the Carnegie-IAS Commission on Mathematics and Science Education. http://www.opportunityequation.org/resources. This online article contains an excellent summary of original, scholarly research examining the theory of malleable intelligence and the effect of mindsets on math and science achievement.

[53] Ericsson, K. A., Charness, N., Feltovich, P. J., & Hoffman, R. R. (Eds.). (2006). *The Cambridge handbook of expertise and expert performance*. New York, NY: Cambridge University Press.

[54] Fryer, R. G. (2006). "Acting white," the social price paid by the best and brightest minority students. *Education Next*, Winter 2006, 53-59.

[55] Gándara, P. (2005). *Fragile futures: Risk and vulnerability among Latino high-achievers*. Princeton, NJ: Policy Evaluation and Research Center, Educational Testing Service. Retrieved from: http://www.ets .org/Media/Research/pdf/PICFRAGFUT.pdf.

[56] Data on Hispanic college-aged student population growth were obtained from the following two sources: 1) National Science Board. (2010). Higher education in science and engineering. In *Science and engineering indicators 2010*. Arlington, VA: National Science Foundation. Retrieved from: http://www.nsf.gov/statistics 2) Hussar, W. J., & Bailey, T. M. (2008). *Projections of education statistics to 2017* (NCES 2008-078). National Center for Education Statistics, Institute of Education Sciences. Washington, DC: U.S. Department of Education. Retrieved from: http://nces.ed.gov/pubs2008/2008078.pdf.

[57] Swanson, J. D. (2006). Breaking through assumptions about low-income, minority gifted students. *Gifted Child Quarterly*, *50*, 11-25. This paper describes the results of a federally funded demonstration project, Project Breakthrough, and provides evidence that rigorous curricular content originally developed for gifted children can increase the achievement of all students and improve the attitudes and expectations of teachers.

[58] Neal, D., & Schanzenbach, D. W. (2007). *Left behind by design: Proficiency counts and test-based accountability* (Working Paper No. 13293). Cambridge, MA: National Bureau of Economic Research. Retrieved from: http://www.nber.org/papers/w13293.pdf.

[59] Some existing programs include: Johns Hopkins University Center for Talented Youth, Illinois Math and Science Academy, Duke Talent Identification Program, Northwestern Center for Talent Development, the FIRST Robotics competition, Math Circles, National Lab Day, and Intel Science Talent Search.

[60] Reis, S. M., Westberg, K. L., Kulikowich, J. M., & Purcell, J. H. (1998). Curriculum compacting and achievement test scores: What does the research say? *Gifted Child Quarterly*, *42*, 123-129.

[61] Ibid.

[62] Colangelo, N., Assouline, S., & Gross, M. U. M. (2004). *A nation deceived: How schools hold back America's brightest students*. Iowa City, IA: The Connie Belin and Jacqueline N. Blank International Center for Gifted Education and Talent Development, University of Iowa. Retrieved from: http://www.acceleratio ninstitute.org/nation_deceived/ND_v1.pdf.

[63] For examples of acceleration interventions, see the following: Institute for Research and Policy on Acceleration National Work Group. (2009). Appendix A: Definitions of acceleration interventions. In *Guidelines for developing an academic acceleration policy* (pp. 12-14). Iowa City, IA: Institute for Research and Policy on Acceleration, Belin-Blank Center for Gifted Education and Talent Development, University of Iowa. Retrieved from:
http://www.accelerationinstitute.org/resources

[64] Swiatek, M. A., & Benbow, C. P. (1992). Nonacademic correlates of satisfaction with accelerative programs. *Journal of Youth and Adolescence*, *21*, 699-723.

[65] For a general discussion of acceleration and its costs and benefits, see: Colangelo, N., Assouline, S., & Gross, M. U. M. (2004). *A nation deceived: How schools hold back America's brightest students*. Iowa City, IA: The Connie Belin and Jacqueline N. Blank International Center for Gifted Education and Talent Development, University of Iowa. Retrieved from: http://www.accelerationinstitute.org/nation_deceived/ND_v1.pdf. The assertion that states and LEAs often deny younger students the opportunity for educational acceleration is supported by the following: The Council of the State Directors of Programs for the Gifted & The National Association for Gifted Children. (2010). *State of the states in gifted education: National policy and practice data 2008-2009*. Washington, DC: Author.

[66] For a discussion of inquiry-based learning and teaching, see: Olson, S., & Loucks-Horsley, S. (Eds.). (2000). *Inquiry and the national science education standards: A guide for teaching and learning*. Committee on the Development of an Addendum to the National Science Education Standards on Scientific Inquiry, National Research Council. Washington, DC: National Academies Press. For a discussion on inquiry-based learning, talent identification and development, and the role of the learning ecosystem in fostering achievement, see: Brandwein, P. F. (1995). *Science talent in the young expressed within ecologies of achievement* (RBDM 9510). Storrs, CT: The National Research Center on the Gifted and Talented, University of Connecticut. Retrieved from: http://www.gifted. For evidence that rigorous, inquiry-based curricula originally developed for gifted children can increase the achievement of all students, see: Swanson, J. D. (2006). Breaking through assumptions about low-income, minority gifted students. *Gifted Child Quarterly*, *50*, 11-25.

[67] For data regarding in-school services provided to talented students and the effects of these services, see the following three references: 1) The Council of the State Directors of Programs for the Gifted & The National Association for Gifted Children. (2010). *State of the states in gifted education: National policy and practice data 2008-2009* (pp. 35-38). Washington, DC: Author. 2) Reis, S. M., McCoach, D. B., Coyne, M., Schreiber, F. J., Eckert, R. D., & Gubbins, E. J. (2007). Using planned enrichment strategies with direct instruction to improve reading fluency, comprehension, and attitude toward reading: An evidence-based study. *The Elementary School Journal*, *108*, 3-24. 3) Gavin, M. K., Casa, T. M., Adelson, J. L., Carroll, S. R., & Shefeld,

L. J. (2009). The impact of advanced curriculum on the achievement of mathematically promising elementary students. *Gifted Child Quarterly*, *53*(3), 188-202.

[68] "Informal Education" refers to a variety of interventions that occur outside of the primary, in-class curriculum. The definition includes, but is not limited to, structured, accelerative summer classes, distance education programs, STEM talent competitions, science fairs, museum visits, and out-of-school laboratory experiences. Informal, in this context, does not necessarily mean "unstructured."

[69] For a discussion on the disparities that exist among students based on income and race/ethnicity, see: 1) Wyner, J. S., Bridgeland, J. M., & DiIulio, J. J. (2007). *Achievement trap: How America is failing millions of high-achieving students from lower-income families*. Landsdowne, VA: Jack Kent Cooke Foundation. Retrieved from: http://www.civicenterprises. net/pdfs/jkc.pdf. 2) Reardon, S. (2008). *Differential growth in the black-white achievement gap during elementary school among high- and low-scoring students* (Working Paper 2008-7). Stanford, CA: Institute for Research on Education Policy & Practice, Stanford University. Retrieved from: http://www.stanford.edu/group/irepp/cgi-bin/joomla/docman/differentialgrowth-in-the-black-white-achievement-gap-reardon/download.html.

[70] For a discussion of enrichment and enrichment in combination with acceleration, see the following three references: 1) Reis, S. M., McCoach, D. B., Coyne, M., Schreiber, F. J., Eckert, R. D., & Gubbins, E. J. (2007). Using planned enrichment strategies with direct instruction to improve reading fluency, comprehension, and attitude toward reading: An evidence-based study. *The Elementary School Journal*, *108*, 3-24. 2) Gavin, M. K., Casa, T. M., Adelson, J. L., Carroll, S. R., & Shefeld, L. J. (2009). The impact of advanced curriculum on the achievement of mathematically promising elementary students. *Gifted Child Quarterly*, *53*(3), 188-202. 3) National Mathematics Advisory Panel. (2008). *Foundations for success: The final report of the National Mathematics Advisory Panel*. Washington, DC: U.S. Department of Education. Retrieved from: http://www2.ed.gov/about/bdscomm/list/mathpanel/report/final-report.pdf.

[71] Tai, R. H., Liu, C. Q., Maltese, A. V., & Fan, X. (2006). Planning early for careers in science. *Science*, *312*, 1143-1144.

[72] Lohman, D. F. (2005). *Identifying academically talented minority students* (RM05216). Storrs, CT: The National Research Center on the Gifted and Talented, University of Connecticut. Retrieved from: http://www.gifted

[73] For a definitive analysis of the research on acceleration, see: Kulik, J. A. (2004). Meta-analytic studies of acceleration. In N. Colangelo, S. Assouline, & M. U. M. Gross (Eds.), *A nation deceived: How schools hold back America's brightest students* (Vol. 2, pp. 13-22). Iowa City, IA: The Connie Belin & Jacqueline N. Blank International Center for Gifted Education and Talent Development, University of Iowa. Retrieved from: http://www.accelerationinstitute.org/nation_deceived/ND_v2.pdf. For a recent examination of the long-term effects of acceleration, enrichment, and other learning opportunities on the achievement of talented students, see: Wai, J., Lubinski, D., Benbow, C. P., & Steiger, J. H. (2010). Accomplishment in science, technology, engineering, and mathematics (STEM) and its relation to STEM educational dose: A 25-year longitudinal study. *Journal of Educational Psychology*, *102* [In Press]. For detailed examples of acceleration policy in practice, see: Institute for Research and Policy on Acceleration National Work Group. (2009). *Guidelines for developing an academic acceleration policy*. Iowa City, IA: Institute for Research and Policy on Acceleration, Belin-Blank Center for Gifted Education and Talent Development, University of Iowa. Retrieved from: http://www. accelerationinstitute.org/resources

[74] For a discussion on the achievement gaps and other educational disparities present among various student sub-groups, see the following three references: 1) Wyner, J. S., Bridgeland, J. M., & DiIulio, J. J. (2007). *Achievement trap: How America is failing millions of high-achieving students from lower-income families*. Landsdowne, VA: Jack Kent Cooke Foundation. Retrieved from: http://www.civicenterprises.net/pdfs/jkc.pdf. 2) Reardon, S. (2008). *Differential growth in the black-white achievement gap during elementary school among high- and low-scoring students* (Working Paper 2008-7). Stanford, CA: Institute for Research on Education Policy & Practice, Stanford University. Retrieved from: http://www. stanford.edu/group/irepp/cgi-bin/joomla/docman/differential-growth-in-the-black-white-achievement-gap-reardon/download. html. 3) Gándara, P. (2005). *Fragile futures: Risk and vulnerability among Latino high-achievers*. Princeton, NJ: Policy Evaluation and Research Center, Educational Testing Service. Retrieved from: http://www.ets.org/Media/Research/pdf/PICFRAGFUT.pdf.

[75] For a discussion on talent development, see: Renzulli, J. S. (2005). The three-ring conception of giftedness: A developmental model for promoting creative productivity. In R. J. Sternberg & J. Davidson (Eds.), *Conceptions of giftedness* (2nd Ed.) (pp. 217-245). Boston, MA: Cambridge University Press. Also, see: Bloom, B. S. (1985). *Developing talent in young people*. New York, NY: Ballantine.

[76] Super, D. E., & Bachrach, P. B. (1957). *Scientific careers and vocational development theory*. New York, NY: Bureau of Publications, Teachers College, Columbia University. For a recent review of multiple longitudinal studies of regarding the role of mathematical and spatial abilities in the development of STEM expertise, see: Wai, J., Lubinski, D., & Benbow, C. P. (2009). Spatial ability for STEM domains: Aligning over 50 years of cumulative psychological knowledge solidifies its importance. *Journal of Educational Psychology*, *101*, 817-835.

[77] Wai, J., Lubinski, D., & Benbow, C. P. (2009). Spatial ability for STEM domains: Aligning over 50 years of cumulative psychological knowledge solidifies its importance. *Journal of Educational Psychology, 101*, 8 17-835.

[78] Lohman, D. F. (2005). An aptitude perspective on talent: Implications for identification of academically gifted minority students. *Journal for the Education of the Gifted, 28*, 333-360.

[79] For a discussion on teachers as "gatekeepers," see: Ford, D. Y., & Grantham, T. C. (2003). Providing access for culturally diverse gifted students: from deficit to dynamic thinking. *Theory Into Practice, 42*, 217-225.

[80] For a discussion on how inadequate teacher training hinders the identification of talented underrepresented minorities, see: Ford, D. Y., & Grantham, T. C. (2003). Providing access for culturally diverse gifted students: from deficit to dynamic thinking. *Theory Into Practice, 42*, 217-225. For a general discussion about the training requirements for teachers in gifted & talented education, see: The Council of the State Directors of Programs for the Gifted & The National Association for Gifted Children. (2010). *State of the states in gifted education: National policy and practice data 2008-2009* (pp. 38-41). Washington, DC: Author.

[81] Grantham, T. C., & Ford, D. Y. (2003). Beyond self-concept and self-esteem for African American students: Improving racial identity improves achievement. *The High School Journal, 87*, 18-29.

[82] The Council of the State Directors of Programs for the Gifted & The National Association for Gifted Children. (2010). *State of the states in gifted education: National policy and practice data 2008-2009* (pp. 38-41). Washington, DC: Author.

[83] For a discussion on how teacher training and experience working with talented students can improve the identification of talented underrepresented minorities, see: Ford, D. Y., & Grantham, T. C. (2003). Providing access for culturally diverse gifted students: from deficit to dynamic thinking. *Theory Into Practice, 42*, 217-225. For a review of the literature concerning the factors that influence the attitudes of educators towards talented students, see: Bégin, J., & Gagné, F. (1994). Predictors of attitudes toward gifted education: A review of the literature and blueprints for future research. *Journal for the Education of the Gifted, 17*, 161-179.

[84] Grantham, T. C., & Ford, D. Y. (2003). Beyond self-concept and self-esteem for African American students: Improving racial identity improves achievement. *The High School Journal, 87*, 18-29.

[85] Beiock, S. L., Gunderson, E. A., Ramirez, G., & Levine, S. C. (2010). Female teachers' math anxiety affects girls' math achievement. *Proceedings of the National Academy of Sciences, 107*, 1860-1863.

[86] Bleske-Rechek, A., Lubinski, D., & Benbow, C. P. (2004). Meeting the educational needs of special populations: Advanced Placement's role in developing exceptional human capital. *Psychological Science, 15*, 217-224.

[87] National Academy of Sciences. (2008). *Changing the conversation: Messages for improving public understanding of engineering*. National Academy of Engineering, Committee on Public Understanding of Engineering Messages. Washington, DC: National Academies Press.

[88] Progress data of the high-achieving cohort should be appropriately disaggregated.

[89] Park, G., Lubinski, D., & Benbow, C. P. (2007). Contrasting intellectual patterns predict creativity in the arts and sciences. *Psychological Science, 18*, 948-9 52.

[90] Roosevelt, F. D. (1945). President Roosevelt's letter. In V. Bush, *Science–the endless frontier. A report to the President on a program for postwar scientific research* (p. 4). Washington, DC: U.S. Government Printing Office.

[91] Bush, V. (1945). Letter of transmittal. In *Science–the endless frontier. A report to the President on a program for postwar scientific research* (p. 2). Washington, DC: U.S. Government Printing Office.

[92] "Innovators" are being defined here as those individuals who have developed the expertise to become leading STEM professionals, and might even have become the creators of significant breakthroughs or advances in scientific and technological understanding - some of which may have completely changed research fields and/or might be patentable, for example.

[93] Recent Board policy reports addressing this subject include the Companion to *Science and Engineering Indicators (SEI) 2008, Research and Development: Essential Foundation for U.S. Competitiveness in a Global Economy* (NSB-08-3), the Companion to *SEI 2006, America's Pressing Challenge—Building a Stronger Foundation*; the Companion to *SEI 2004, An Emerging and Critical Problem of the Science and Engineering Labor Force* (NSB-04-7), *Moving Forward to Improve Engineering Education* (NSB-07-122), and *The Science and Engineering Workforce—Realizing America's Potential* (NSB-04-69).

[94] *Final Report of the National Mathematics Advisory Panel*, March 2008, Department of Education.

Chapter Sources

The following chapters have been previously published:

Chapter 1 – This is an edited, excerpted and augmented edition of a United States President's Council of Advisors on Science and Technology publication, dated September 2010.

Chapter 2 – This is an edited, excerpted and augmented edition of a National Science Foundation (National Science Board) publication, dated May 5, 2010.

INDEX

C

D

E

N

O

S

V

W

Y